MATH WORKBOOK

4TH GRADE

1,500+ Practice Questions

MATH PRACTICE QUESTIONS FOR DAILY EXERCISE

Written and edited by APEX Test Prep.

ISBN 13: 978-1-6284-5882-4
ISBN 10: 1-6284-5882-8

For additional information or for bulk orders, contact info@apexprep.com.

Introduction

This workbook was developed to help 4th grade students master all of the important math skills that they have learned up to this point. It starts with the most basic concepts and works up to some rather advanced concepts. It is broken down into the four main content areas listed below:

- Arithmetic
- Measurement and Data
- Algebra
- Geometry

While this workbook is filled with over 1,500 practice problems, it may not always have enough room to work the problems. We recommend that students keep some scratch paper nearby to help them solve some of the problems.

Our goal in creating this workbook was to help 4th grade students get plenty of practice across all areas of math. With that being said, if you found something that you felt wasn't up to your standards, then please send us an email at the address below.

Thanks Again and Happy Studying!
APEX Test Prep
info@apexprep.com

Table of Contents

Table of Contents

Table of Contents

Table of Contents

Arithmetic

Identifying Place Value

Identify the place value of the underlined digit in each number.

1 3,5__3__5

_____

2 9__4__,502

_____

3 __2__0,031

_____

4 19,__4__85

_____

5 539,11__8__

_____

6 __3__44,106

_____

7 512,__2__1

_____

8 946,7__0__6

_____

9 6__4__9,929

_____

10 432,45__2__

_____

11 5__2__1,853

_____

12 __1__34,223

_____

13 215,__8__92

_____

14 __9__,129,938

_____

15 12,3__4__2,934

_____

Identifying the Place Value of a Given Number

Identify the place value of the given number.

1 854,611 - The 6

_____

2 672,589 - The 8

_____

3 109,204 - The 1

_____

4 515,650 - The 0

_____

5 331,900 - The 9

_____

6 583,540 - The 8

_____

7 86,913 - The 6

_____

8 589,426 - The 2

_____

9 6,288,790 - The 6

_____

10 16,682 - The 1

_____

11 5,509.40 - The 4

_____

12 156,202 - The 1

_____

13 475,681 - The 7

_____

14 9.521 - The 9

_____

15 697.751 - The 5

_____

Compare the Values

Compare the value of the underlined digit given its place,
not the value of the entire integer. Use the symbols =, <, or >.
The first one has been completed as an example.

1

3<u>2</u>75 __<__ 2<u>6</u>52

2

765<u>6</u> _____ 183<u>3</u>

3

4<u>5</u>834 _____ 9<u>5</u>384

4

745<u>7</u>1 _____ 1<u>1</u>743

5

65<u>0</u>346 _____ <u>7</u>250

6

5.656<u>7</u> _____ 5287.<u>7</u>

7

565.7<u>1</u>21 _____ 163.<u>4</u>539

8

2<u>0</u>.740 _____ 896.<u>5</u>20

9

354.1<u>7</u>9 _____ 90324.<u>1</u>

10

415.7<u>9</u>52 _____ 7724.4<u>5</u>

Decomposing Numbers

Decompose each number by writing how many hundreds, tens, ones, etc. the number has.

1 798

___ ten thousands, ___ thousands, ___ hundreds, ___ tens, ___ ones, ___ tenths, ___ hundredths, ___ thousandths, ___ ten thousandths

2 501

___ hundred thousands, ___ ten thousands, ___ thousands, ___ hundreds, ___ tens, ___ ones, ___ tenths, ___ hundredths, ___ thousandths, ___ ten thousandths

3 927.05

___ hundred thousands, ___ ten thousands, ___ thousands, ___ hundreds, ___ tens, ___ ones, ___ tenths, ___ hundredths, ___ thousandths, ___ ten thousandths

4 260.23

___ hundred thousands, ___ ten thousands, ___ thousands, ___ hundreds, ___ tens, ___ ones, ___ tenths, ___ hundredths, ___ thousandths, ___ ten thousandths

5 2,709

___ hundred thousands, ___ ten thousands, ___ thousands, ___ hundreds, ___ tens, ___ ones, ___ tenths, ___ hundredths, ___ thousandths, ___ ten thousandths

6 6,813

___ hundred thousands, ___ ten thousands, ___ thousands, ___ hundreds, ___ tens, ___ ones, ___ tenths, ___ hundredths, ___ thousandths, ___ ten thousandths

7 1,982

___ hundred thousands, ___ ten thousands, ___ thousands, ___ hundreds, ___ tens, ___ ones, ___ tenths, ___ hundredths, ___ thousandths, ___ ten thousandths

8 8,714

___ hundred thousands, ___ ten thousands, ___ thousands, ___ hundreds, ___ tens, ___ ones, ___ tenths, ___ hundredths, ___ thousandths, ___ ten thousandths

Decomposing Numbers

Decompose each number by writing how many hundreds, tens, ones, etc. the number has.

9) 11,920

____ hundred thousands, ____ ten thousands, ____ thousands, ____ hundreds, ____ tens, ____ ones, ____ tenths, ____ hundredths, ____ thousandths, ____ ten thousandths

10) 90,101

____ hundred thousands, ____ ten thousands, ____ thousands, ____ hundreds, ____ tens, ____ ones, ____ tenths, ____ hundredths, ____ thousandths, ____ ten thousandths

11) 45,341

____ hundred thousands, ____ ten thousands, ____ thousands, ____ hundreds, ____ tens, ____ ones, ____ tenths, ____ hundredths, ____ thousandths, ____ ten thousandths

12) 29,800

____ hundred thousands, ____ ten thousands, ____ thousands, ____ hundreds, ____ tens, ____ ones, ____ tenths, ____ hundredths, ____ thousandths, ____ ten thousandths

13) 100,954

____ hundred thousands, ____ ten thousands, ____ thousands, ____ hundreds, ____ tens, ____ ones, ____ tenths, ____ hundredths, ____ thousandths, ____ ten thousandths

14) 302,540

____ hundred thousands, ____ ten thousands, ____ thousands, ____ hundreds, ____ tens, ____ ones, ____ tenths, ____ hundredths, ____ thousandths, ____ ten thousandths

15) 854,624

____ hundred thousands, ____ ten thousands, ____ thousands, ____ hundreds, ____ tens, ____ ones, ____ tenths, ____ hundredths, ____ thousandths, ____ ten thousandths

Writing Numbers From Their Components

Write the number formed by composing the number of hundreds, tens, ones, etc. indicated in the problem.

1 600 + 70 + 2

_____

2 700 + 60

_____

3 100 + 50 + 3

_____

4 1000 + 2

_____

5 1000 + 400 + 30 + 9

_____

6 5 hundreds, 8 tens, 2 ones

_____

7 3 hundreds, 1 ten, 9 ones

_____

8 7 hundreds, 0 tens, 4 ones

_____

9 1 hundred, 2 tens, 6 ones

_____

10 8 hundreds, 7 tens, 6 ones

_____

11 9 hundreds, 4 tens, 0 ones, 1 tenths, 3 hundredths

_____

12 4 hundreds, 9 tens, 7 ones, 0 tenths, 0 hundredths, 4 thousandths

_____

13 8 thousands, 2 hundreds, 0 tens, 1 ones

_____

14 6 hundreds, 5 tens, 8 ones, 9 tenths

_____

15 3 hundreds

_____

16 1 ten thousands, 4 thousands, 9 hundreds, 4 tens, 1 one

_____

17 8 ten thousands, 0 thousands, 6 hundreds, 6 ten, 7 ones

_____

18 1 tenth, 8 hundredths, 4 thousandths

_____

19 8 hundreds, 1 ten, 0 ones

_____

20 4 hundreds, 7 tens, 5 ones

_____

Reading and Writing Numbers

For the first 10 problems, use the provided line to write out how the number next to it is read. The first one has been done as an example. For the 15 problems that follow, write the number indicated by the words given.

1 9747

Nine thousand seven hundred forty-seven

2 639,032

3 81,300

4 2,324,371

5 801,690,802

6 5394.63

7 120,638.5

8 7,470.003

Reading and Writing Numbers

For the first 10 problems, use the provided line to write out how the number next to it is read. For the 15 problems that follow, write the number indicated by the words given.

9 97,009.7111

_____

10 0.08

_____

11 One thousand

_____

12 Fifty-seven

_____

13 One hundred twelve

_____

14 Eight hundred

_____

15 Four hundred seventy-five

_____

16 Seven hundred two

_____

Reading and Writing Numbers

For the first 10 problems, use the provided line to write out how the number next to it is read. For the 15 problems that follow, write the number indicated by the words given.

17   **Five hundred eighty**

_____

18   **Nine hundred seventy-six**

_____

19   **Two hundred ninety**

_____

20   **Three hundred twenty-nine**

_____

21   **Seven hundred one**

_____

22   **Six hundred sixteen**

_____

23   **Eight hundred eighty-eight**

_____

24   **Four hundred nine**

_____

25   **Nine hundred thirty-two**

_____

Practice Makes Perfect: Addition and Subtraction Review

Compute the sums or differences as stated in the problems below.

1) $76 + 84$

= _____

2) $99 + 15$

= _____

3) $29 + 74$

= _____

4) $976 + 566$

= _____

5) $386 + 985$

= _____

6) $649 + 686$

= _____

7) $858 + 893$

= _____

8) $9938 + 1095$

= _____

9) $5889 + 8215$

= _____

10) $8835 + 1995$

= _____

11) $178 - 89$

= _____

12) $73 - 41 =$

_____

13) $901 - 68$

= _____

14) $260 - 72$

= _____

15) $7720 - 868$

= _____

16) $9102 - 645$

= _____

17) $5800 - 833$

= _____

18) $951 - 620$

= _____

19) $55218 - 9739$

= _____

20) $25460 - 9568$

= _____

21) $638 + 84$

= _____

22) $519 + 69$

= _____

23) $119 + 63$

= _____

24) $201 + 21$

= _____

25) $853 + 10$

= _____

26) $468 + 39$

= _____

27) $545 + 43$

= _____

28) $402 + 34$

= _____

29) $120 + 68$

= _____

30) $300 + 35$

= _____

31) $377 - 73$

= _____

32) $880 - 22$

= _____

33) $478 - 38$

= _____

34) $495 - 21$

= _____

35) $249 - 17$

= _____

36) $698 - 57$

= _____

Practice Makes Perfect: Addition and Subtraction Review

Compute the sums or differences as stated in the problems below.

37
468
- 41
_____

38
743
- 38
_____

39
231
- 26
_____

40
256
- 25
_____

41
304
- 71
_____

42
367
- 45
_____

43
826
- 39
_____

44
337
- 62
_____

45
376
- 69
_____

46
893
- 94
_____

47
781
- 84
_____

48
412
- 62
_____

49
644
- 83
_____

50
421
- 38
_____

51
993
- 10
_____

52
822
- 42
_____

53
595
- 86
_____

54
203
- 25
_____

55
433
- 77
_____

56
516
- 45
_____

57
162
- 46
_____

58
333
- 47
_____

59
103
- 77
_____

60
323
- 56
_____

61
543
- 97
_____

62
711
- 75
_____

63
770
- 34
_____

64
465
- 49
_____

65
182
- 19
_____

66
899
- 87
_____

67
296
- 57
_____

68
452
- 49
_____

69
832
- 40
_____

70
954
- 59
_____

Real-World Addition and Subtraction Problems

1 Fifteen dogs are at the shelter waiting for homes. There are also seven cats. How many animals are there in total?

2 Li Wei's dad is on a diet. His doctor wants him to lose 31 pounds. If he has lost 8 pounds in the first month and 7 pounds over the next month, how many pounds does he have left to lose?

3 Meng and her mom are making steamed dumplings. The first batch had 9 dumplings, and the second batch also had 9. How many have they made in total?

4 A house has 6 rooms downstairs and 5 upstairs. How many rooms are there all together?

5 Derek has 8 markers and 6 crayons on his table. How many drawing instruments does he have?

6 Tommy has 2 dogs, 1 rabbit, and 4 goldfish. How many pets does he have?

7 Shayna has 2 grandpas, 2 grandmas, 3 aunts, 4 uncles, and 3 cousins. How many extended family members does she have?

8 Peter has 2 basketballs, 1 baseball, 5 tennis balls, 4 footballs, and 1 football. How many balls does he have in total?

9 It is five weeks into summer vacation. If there are four weeks left until school starts, how many weeks was summer break?

10 Jada made 20 friendship bracelets. If she gave 5 to her family members and 7 to friends, how many does she have left?

Real-World Addition and Subtraction Problems

11 A first grade teacher has 23 students. 12 are boys. How many are girls?

12 A movie theater is showing 4 action films, 3 kids' movies, 5 comedies, and 2 documentaries. How many films are showing?

13 Kyle went to the library and checked out 3 movies, 6 picture books, and 3 audiobooks. How many items did he get?

14 You are washing dishes after family dinner. How many dishes must you wash all together if there are 5 plates, 5 bowls, 5 glasses, and 2 salad bowls?

15 A football team buys 8 cheese pizzas, 3 pepperoni pizzas, 5 vegetable pizzas, and 2 Hawaiian pizzas. How many did they get?

16 A dance team is having a bagel breakfast. 4 kids want plain bagels, 8 kids want poppy seed bagels, 3 kids want sesame seed, and 2 want cinnamon raisin bagels. How many are purchased all together?

17 Felicity is making gifts for seniors in the nursing home. She knits 4 scarves, 5 lap blankets, and 6 hats. How many gifts did she knit?

18 Jackson has to practice piano 45 minutes after school each day. He has played for 27 minutes so far today. How much longer does he need to practice?

19 Your dad is helping you bake cookies for a fundraiser for your soccer team. You volunteered to bake 50 cookies. So far, you have baked 34 cookies. How many do you have left to make?

20 Teddy is selling raffle tickets to raise money for his basketball team. His coach asked each athlete to sell 20 tickets. Teddy has sold 13 so far. How many tickets does he still need to sell?

Real-World Addition and Subtraction Problems

21 Kareem had seven action figures. He then gets four more at a yard sale. How many action figures does he have now?

22 Freddy got to the movie 35 minutes early. He waited in line for tickets for 12 minutes and then waited for popcorn for 5 minutes. How much time is left before the movie?

23 Nelly's stepmom is a car mechanic. If she repaired 16 cars today and five of those were before lunch, how many cars did she work on in the afternoon?

24 Chantal had 42 raisins in her box and ate 15. How many does she have left?

25 Ms. Smith asks his first graders what their favorite toy is. 7 students like playing with Legos, 7 like dolls, 3 like animal figures, and 6 like toy vehicles. How many students are in his class?

26 An elementary school orchestra has 16 violin players, 5 viola players, 12 cello players, and 2 upright bass players. How many students are in the orchestra?

27 A farmer has 7 cows, 5 goats, 5 sheep, 12 chickens, and 1 horse. How many animals are on the farm?

28 After school, Benny eats 12 carrot sticks, 6 slices of cheese, and 12 grapes. How many items is this?

29 If Jonah is allowed to watch 30 minutes of TV per night and has watched 15 minutes so far, how much longer can he watch TV today?

30 You got a $25 gift card to an art store. You bought a paint set for $14 and brushes for $4. How much do you have left on your gift card?

Multiplying a 1-Digit Number by a 1-Digit Number

1 6 x 3 = _____

2 5 x 7 = _____

3 3 x 8 = _____

4 6 x 4 = _____

5 4 x 8 = _____

6 1 x 6 = _____

7 9 x 7 = _____

8 0 x 4 = _____

9 8 x 1 = _____

10 5 x 3 = _____

11 8 x 5 = _____

12 4 x 6 = _____

13 9 x 3 = _____

14 6 x 2 = _____

15 1 x 2 = _____

16 0 x 6 = _____

17 7 x 3 = _____

18 4 x 5 = _____

19 5 x 1 = _____

20 2 x 9 = _____

21 1 x 5 = _____

22 9 x 9 = _____

23 0 x 5 = _____

24 7 x 7 = _____

25 6 x 8 = _____

Multiplying a 2-Digit Number by a 1-Digit Number

1 16 x 5 = _____

2 24 x 8 = _____

3 50 x 9 = _____

4 36 x 1 = _____

5 56 x 7 = _____

6 64 x 9 = _____

7 86 x 4 = _____

8 59 x 7 = _____

9 40 x 0 = _____

10 92 x 1 = _____

11 73 x 9 = _____

12 67 x 6 = _____

13 44 x 9 = _____

14 72 x 3 = _____

15 34 x 0 = _____

16 74 x 9 = _____

17 61 x 0 = _____

18 68 x 6 = _____

19 93 x 6 = _____

20 76 x 4 = _____

21 51 x 8 = _____

22 80 x 8 = _____

23 11 x 2 = _____

24 29 x 1 = _____

25 17 x 0 = _____

Multiplying a 2-Digit Number by a 2-Digit Number

1) 65
x 76
= _____

2) 20
x 43
= _____

3) 61
x 78
= _____

4) 99
x 16
= _____

5) 61
x 78
= _____

6) 26
x 54
= _____

7) 55
x 87
= _____

8) 47
x 19
= _____

9) 18
x 66
= _____

10) 54
x 27
= _____

11) 15
x 66
= _____

12) 99
x 79
= _____

13) 67
x 15
= _____

14) 17
x 44
= _____

15) 66
x 77
= _____

16) 51
x 50
= _____

17) 79
x 39
= _____

18) 15
x 48
= _____

19) 38
x 12
= _____

20) 34
x 94
= _____

21) 45
x 49
= _____

22) 30
x 70
= _____

23) 24
x 49
= _____

24) 76
x 34
= _____

25) 48
x 57
= _____

Multiplying a 3-Digit Number by a 2-Digit Number

1) 465
x 96
= _____

2) 867
x 89
= _____

3) 108
x 57
= _____

4) 453
x 68
= _____

5) 566
x 99
= _____

6) 288
x 21
= _____

7) 975
x 17
= _____

8) 536
x 76
= _____

9) 376
x 78
= _____

10) 606
x 26
= _____

11) 299
x 40
= _____

12) 446
x 31
= _____

13) 264
x 39
= _____

14) 333
x 24
= _____

15) 187
x 26
= _____

16) 635
x 39
= _____

17) 465
x 66
= _____

18) 874
x 40
= _____

19) 638
x 52
= _____

20) 511
x 69
= _____

21) 733
x 31
= _____

22) 858
x 43
= _____

23) 114
x 98
= _____

24) 792
x 49
= _____

25) 388
x 55
= _____

Dividing by 1 – 10

1   15 ÷ 5 = _____

2   60 ÷ 6 = _____

3   72 ÷ 9 = _____

4   70 ÷ 10 = _____

5   6 ÷ 1 = _____

6   8 ÷ 2 = _____

7   30 ÷ 6 = _____

8   56 ÷ 8 = _____

9   9 ÷ 3 = _____

10   27 ÷ 9 = _____

11   49 ÷ 7 = _____

12   3 ÷ 3 = _____

13   18 ÷ 2 = _____

14   30 ÷ 6 = _____

15   4 ÷ 1 = _____

16   42 ÷ 1 = _____

17   30 ÷ 3 = _____

18   4 ÷ 2 = _____

19   90 ÷ 9 = _____

20   6 ÷ 3 = _____

21   30 ÷ 5 = _____

22   12 ÷ 4 = _____

23   60 ÷ 6 = _____

24   12 ÷ 3 = _____

25   15 ÷ 3 = _____

Dividing by 1 – 100

1) $3258 \div 36 =$ _____

2) $8300 \div 83 =$ _____

3) $5544 \div 84 =$ _____

4) $855 \div 9 =$ _____

5) $29 \div 58 =$ _____

6) $121 \div 11 =$ _____

7) $837 \div 31 =$ _____

8) $1026 \div 57 =$ _____

9) $1972 \div 58 =$ _____

10) $2500 \div 50 =$ _____

11) $4187 \div 53 =$ _____

12) $2772 \div 66 =$ _____

13) $4692 \div 68 =$ _____

14) $4092 \div 44 =$ _____

15) $168 \div 3 =$ _____

16) $2376 \div 66 =$ _____

17) $3901 \div 47 =$ _____

18) $783 \div 9 =$ _____

19) $602 \div 86 =$ _____

20) $860 \div 43 =$ _____

21) $3410 \div 62 =$ _____

22) $59 \div 59 =$ _____

23) $4056 \div 78 =$ _____

24) $3658 \div 59 =$ _____

25) $4070 \div 55 =$ _____

Dividing a 3-Digit by 1-Digit and 2-Digit Numbers

1) 473 ÷ 43 = _____

2) 760 ÷ 8 = _____

3) 455 ÷ 65 = _____

4) 609 ÷ 87 = _____

5) 455 ÷ 35 = _____

6) 344 ÷ 43 = _____

7) 648 ÷ 8 = _____

8) 936 ÷ 13 = _____

9) 765 ÷ 85 = _____

10) 693 ÷ 77 = _____

11) 810 ÷ 10 = _____

12) 403 ÷ 31 = _____

13) 902 ÷ 82 = _____

14) 308 ÷ 77 = _____

15) 882 ÷ 49 = _____

16) 800 ÷ 40 = _____

17) 738 ÷ 9 = _____

18) 592 ÷ 16 = _____

19) 741 ÷ 57 = _____

20) 980 ÷ 70 = _____

21) 426 ÷ 71 = _____

22) 901 ÷ 53 = _____

23) 712 ÷ 89 = _____

24) 756 ÷ 36 = _____

25) 704 ÷ 88 = _____

Prime or Composite?

Determine whether each number below is a prime or composite number. Circle P if the number is a prime number or circle C if the number is a composite number.

1. 5
P   C

2. 30
P   C

3. 6
P   C

4. 13
P   C

5. 74
P   C

6. 91
P   C

7. 88
P   C

8. 17
P   C

9. 20
P   C

10. 81
P   C

11. 71
P   C

12. 89
P   C

13. 33
P   C

14. 73
P   C

15. 47
P   C

Greatest Common Factor

Determine the greatest common factor for each pair of numbers.

1 30 and 5

_____

2 60 and 2

_____

3 10 and 6

_____

4 6 and 5

_____

5 15 and 6

_____

6 60 and 12

_____

7 5 and 24

_____

8 20 and 30

_____

9 4 and 60

_____

10 12 and 15

_____

11 4 and 20

_____

12 30 and 40

_____

13 44 and 55

_____

14 39 and 91

_____

15 96 and 80

_____

Least Common Multiple

Determine the least common multiple for each pair of numbers.

1 40 and 6

_____

2 30 and 2

_____

3 3 and 12

_____

4 3 and 15

_____

5 5 and 12

_____

6 4 and 60

_____

7 10 and 6

_____

8 2 and 3

_____

9 40 and 20

_____

10 5 and 24

_____

11 12 and 24

_____

12 4 and 40

_____

13 3 and 5

_____

14 5 and 30

_____

15 12 and 30

_____

Prime Factorization

Fill in the prime factorization trees where displayed or list the prime factorization for each number if just a number is given.

1

18

18 = ___ X ___ X ___

2

24

24 = ___ X ___ X ___ X ___

3

36

36 = ___ X ___ X ___ X ___

4

54

54 = ___ X ___ X ___ X ___

5

126

126 = ___ X ___ X ___ X ___

6

108

108 = ___ X ___ X ___ X ___ X ___

7

51 = _____

8

44 = _____

9

85 = _____

10

46 = _____

11

74 = _____

12

90 = _____

13

48 = _____

14

94 = _____

15

128 = _____

Fill in the Table

Complete the table by rounding to the nearest value indicated by each column.
The first one has been completed as an example.

	Hundredths	Tenths	Ones	Tens	Hundreds	Thousands
8316.728	8316.73	8316.7	8318	8320	8300	8000
528.652						
4381.748						
1273.025						
9924.703						
3092.816						
501.105						
9281.002						
278.056						
2244.525						
735.642						
1387.955						
984.058						
3629.765						
7445.869						
8401.836						
5832.547						
675.498						
3268.027						
499.367						

Using Rounding to Estimate the Sum or Difference

Round each number to the nearest 10 to estimate the sum or difference.
Write the equation of the rounded numbers and then solve it.
The first one has been done for you as an example.

1 54 + 38

50 + 40 = 90

2 78 − 42

_____

3 17 + 27

_____

4 92 − 24

_____

5 68 + 78

_____

6 43 − 12

_____

7 105 + 59

_____

8 244 − 77

_____

9 339 + 586

_____

10 327 + 316

_____

11 896 − 528

_____

12 886 + 354

_____

13 696 + 552

_____

14 858 + 879

_____

15 912 + 875

_____

Using Rounding to Estimate the Product or Quotient

Round each number to the nearest 10 to estimate the product or quotient.
Write the equation of the rounded numbers and then solve it.
The first one has been done for you as an example.

1 9 x 39

10 x 40 = 400
_____

2 18 x 29

_____

3 43 x 78

_____

4 92 x 77

_____

5 71 x 43

_____

6 26 x 13

_____

7 83 x 68

_____

8 51 x 38

_____

9 15 x 26

_____

10 88 x 39

_____

11 94 x 35

_____

12 61 x 83

_____

13 22 x 58

_____

14 78 ÷ 19

_____

15 99 ÷ 22

_____

Reducing Fractions

Reduce the following fractions to their simplest form.

1. $\dfrac{24}{40}$ = _____

2. $\dfrac{21}{28}$ = _____

3. $\dfrac{8}{24}$ = _____

4. $\dfrac{9}{18}$ = _____

5. $\dfrac{21}{30}$ = _____

6. $\dfrac{27}{63}$ = _____

7. $\dfrac{3}{15}$ = _____

8. $\dfrac{5}{10}$ = _____

9. $\dfrac{25}{50}$ = _____

10. $\dfrac{16}{48}$ = _____

11. $\dfrac{3}{6}$ = _____

12. $\dfrac{27}{63}$ = _____

13. $\dfrac{8}{12}$ = _____

14. $\dfrac{4}{20}$ = _____

15. $\dfrac{6}{12}$ = _____

16. $\dfrac{72}{81}$ = _____

17. $\dfrac{14}{49}$ = _____

18. $\dfrac{6}{24}$ = _____

19. $\dfrac{20}{30}$ = _____

20. $\dfrac{21}{49}$ = _____

21. $\dfrac{36}{40}$ = _____

22. $\dfrac{16}{32}$ = _____

23. $\dfrac{18}{81}$ = _____

24. $\dfrac{24}{64}$ = _____

25. $\dfrac{60}{72}$ = _____

Reducing Fractions

Reduce the following fractions to their simplest form.

1) $\frac{10}{80}$ = _____

2) $\frac{12}{32}$ = _____

3) $\frac{14}{77}$ = _____

4) $\frac{10}{14}$ = _____

5) $\frac{15}{55}$ = _____

6) $\frac{7}{21}$ = _____

7) $\frac{8}{72}$ = _____

8) $\frac{18}{27}$ = _____

9) $\frac{42}{63}$ = _____

10) $\frac{12}{15}$ = _____

11) $\frac{16}{80}$ = _____

12) $\frac{10}{25}$ = _____

13) $\frac{28}{48}$ = _____

14) $\frac{90}{108}$ = _____

15) $\frac{90}{110}$ = _____

16) $\frac{72}{90}$ = _____

17) $\frac{16}{36}$ = _____

18) $\frac{21}{42}$ = _____

19) $\frac{5}{45}$ = _____

20) $\frac{27}{36}$ = _____

21) $\frac{15}{18}$ = _____

22) $\frac{72}{81}$ = _____

23) $\frac{15}{35}$ = _____

24) $\frac{30}{33}$ = _____

25) $\frac{24}{40}$ = _____

Comparing Fractions

A soccer team is having a bake sale. At the end of the day, the fraction remaining of different treats are compared. For each pair of treats listed, fill in the appropriate symbol (<,≤,>,≥,or=) on each line to make each statement true.

1 3/8 of an apple pie _____ 2/7 of a cherry pie

2 3/9 of a lemon meringue pie _____ 1/2 of a pecan pie

3 4/5 of a pumpkin pie _____ 2/6 of a blueberry pie

4 5/6 of a chocolate crème pie _____ 2/4 of a berry crumble

5 6/9 of a walnut cake _____ 2/3 of the vanilla cupcakes

6 6/7 of the lemon squares _____ 5/6 of the pineapple upside-down cake

7 4/12 of the zucchini bread _____ 1/3 of the cinnamon bread

8 2/3 of the chocolate chip cookies _____ 5/9 of the peanut butter balls

9 6/8 of the chocolate fudge _____ 7/9 of the popcorn balls

10 1/3 of the blueberry muffins _____ 1/4 of the corn muffins

Equivalent Fractions

Fill in the missing numerators and denominators to complete the equivalent fractions.

1 $\dfrac{1}{2} = \dfrac{3}{} = \dfrac{5}{} = \dfrac{2}{} = \dfrac{7}{} = \dfrac{4}{} = \dfrac{6}{}$

2 $\dfrac{28}{32} = \dfrac{}{56} = \dfrac{}{24} = \dfrac{42}{40} = \dfrac{}{40} = \dfrac{14}{} = \dfrac{7}{}$

3 $\dfrac{20}{24} = \dfrac{}{6} = \dfrac{15}{} = \dfrac{25}{} = \dfrac{}{42} = \dfrac{}{12} = \dfrac{}{36}$

4 $\dfrac{2}{3} = \dfrac{}{21} = \dfrac{}{9} = \dfrac{12}{} = \dfrac{}{6} = \dfrac{8}{} = \dfrac{}{15}$

5 $\dfrac{4}{16} = \dfrac{}{8} = \dfrac{}{28} = \dfrac{6}{} = \dfrac{}{12} = \dfrac{1}{} = \dfrac{5}{}$

6 $\dfrac{14}{18} = \dfrac{}{63} = \dfrac{35}{} = \dfrac{42}{} = \dfrac{}{9} = \dfrac{}{36} = \dfrac{}{27}$

7 $\dfrac{3}{15} = \dfrac{}{10} = \dfrac{}{35} = \dfrac{4}{} = \dfrac{}{30} = \dfrac{5}{} = \dfrac{1}{}$

8 $\dfrac{63}{70} = \dfrac{}{10} = \dfrac{27}{} = \dfrac{36}{} = \dfrac{}{50} = \dfrac{}{60} = \dfrac{}{20}$

9 $\dfrac{9}{10} = \dfrac{}{60} = \dfrac{63}{} = \dfrac{36}{} = \dfrac{}{50} = \dfrac{}{30} = \dfrac{}{20}$

10 $\dfrac{6}{18} = \dfrac{}{12} = \dfrac{3}{} = \dfrac{7}{} = \dfrac{}{3} = \dfrac{}{6} = \dfrac{}{15}$

Adding Fractions

1 $\dfrac{1}{5} + \dfrac{1}{3}$

= _____

2 $\dfrac{1}{6} + \dfrac{2}{3}$

= _____

3 $\dfrac{7}{8} + \dfrac{3}{4}$

= _____

4 $\dfrac{2}{4} + \dfrac{1}{3}$

= _____

5 $\dfrac{13}{18} + \dfrac{5}{54}$

= _____

6 $\dfrac{3}{4} + \dfrac{3}{10}$

= _____

7 $\dfrac{1}{2} + \dfrac{3}{5}$

= _____

8 $\dfrac{1}{3} + \dfrac{1}{4}$

= _____

9 $\dfrac{1}{5} + \dfrac{3}{4}$

= _____

10 $\dfrac{2}{3} + \dfrac{2}{16}$

= _____

11 $\dfrac{1}{4} + \dfrac{5}{9}$

= _____

12 $\dfrac{2}{5} + \dfrac{1}{4}$

= _____

13 $\dfrac{1}{5} + \dfrac{2}{3}$

= _____

14 $\dfrac{1}{4} + \dfrac{18}{46}$

= _____

15 $\dfrac{3}{10} + \dfrac{9}{15}$

= _____

Subtracting Fractions

1 $\dfrac{2}{4} - \dfrac{1}{2}$
= _____

2 $\dfrac{3}{4} - \dfrac{1}{10}$
= _____

3 $\dfrac{3}{5} - \dfrac{1}{4}$
= _____

4 $\dfrac{1}{2} - \dfrac{1}{4}$
= _____

5 $\dfrac{1}{4} - \dfrac{1}{5}$
= _____

6 $\dfrac{7}{10} - \dfrac{3}{5}$
= _____

7 $\dfrac{2}{4} - \dfrac{1}{3}$
= _____

8 $\dfrac{2}{3} - \dfrac{1}{2}$
= _____

9 $\dfrac{8}{10} - \dfrac{1}{3}$
= _____

10 $\dfrac{3}{4} - \dfrac{3}{10}$
= _____

11 $\dfrac{2}{4} - \dfrac{2}{10}$
= _____

12 $\dfrac{3}{4} - \dfrac{1}{2}$
= _____

13 $\dfrac{1}{2} - \dfrac{1}{8}$
= _____

14 $\dfrac{1}{2} - \dfrac{1}{5}$
= _____

15 $\dfrac{3}{4} - \dfrac{1}{4}$
= _____

Subtracting Mixed Numbers

1 $7\frac{4}{30} - 3\frac{8}{60}$

= _____

2 $9\frac{1}{2} - 1\frac{2}{4}$

= _____

3 $9\frac{1}{2} - 1\frac{3}{4}$

= _____

4 $7\frac{6}{14} - 2\frac{11}{28}$

= _____

5 $8\frac{15}{25} - 2\frac{7}{10}$

= _____

6 $7\frac{4}{10} - 3\frac{3}{4}$

= _____

7 $6\frac{5}{10} - 3\frac{2}{3}$

= _____

8 $6\frac{1}{8} - 4\frac{1}{3}$

= _____

9 $9\frac{2}{5} - 1\frac{1}{2}$

= _____

10 $7\frac{1}{4} - 2\frac{4}{5}$

= _____

11 $8\frac{1}{3} - 1\frac{2}{4}$

= _____

12 $7\frac{1}{4} - 2\frac{1}{3}$

= _____

13 $6\frac{2}{3} - 2\frac{3}{4}$

= _____

14 $7\frac{1}{4} - 2\frac{1}{3}$

= _____

15 $6\frac{7}{10} - 3\frac{4}{5}$

= _____

16 $5\frac{2}{3} - 3\frac{7}{10}$

= _____

17 $7\frac{1}{3} - 4\frac{1}{2}$

= _____

18 $5\frac{1}{5} - 2\frac{1}{3}$

= _____

19 $8\frac{2}{10} - 4\frac{1}{2}$

= _____

20 $5\frac{2}{5} - 1\frac{1}{2}$

= _____

Multiplying Fractions

1) $\dfrac{1}{3} \times \dfrac{4}{8}$
= _____

2) $\dfrac{1}{2} \times \dfrac{6}{8}$
= _____

3) $\dfrac{3}{6} \times \dfrac{3}{10}$
= _____

4) $\dfrac{1}{5} \times \dfrac{2}{4}$
= _____

5) $\dfrac{4}{5} \times \dfrac{4}{7}$
= _____

6) $\dfrac{1}{4} \times \dfrac{1}{3}$
= _____

7) $\dfrac{5}{7} \times \dfrac{8}{9}$
= _____

8) $\dfrac{5}{6} \times \dfrac{2}{4}$
= _____

9) $\dfrac{1}{2} \times \dfrac{5}{8}$
= _____

10) $\dfrac{1}{9} \times \dfrac{3}{5}$
= _____

11) $\dfrac{1}{7} \times \dfrac{1}{5}$
= _____

12) $\dfrac{2}{18} \times \dfrac{6}{8}$
= _____

13) $\dfrac{3}{7} \times \dfrac{3}{9}$
= _____

14) $\dfrac{2}{5} \times \dfrac{1}{10}$
= _____

15) $\dfrac{2}{8} \times \dfrac{3}{9}$
= _____

16) $\dfrac{3}{15} \times \dfrac{3}{8}$
= _____

17) $\dfrac{4}{5} \times \dfrac{1}{4}$
= _____

18) $\dfrac{5}{9} \times \dfrac{1}{3}$
= _____

19) $\dfrac{2}{6} \times \dfrac{8}{9}$
= _____

20) $\dfrac{4}{7} \times \dfrac{1}{2}$
= _____

21) $\dfrac{5}{6} \times \dfrac{4}{5}$
= _____

22) $\dfrac{8}{10} \times \dfrac{1}{7}$
= _____

23) $\dfrac{6}{10} \times \dfrac{5}{6}$
= _____

24) $\dfrac{1}{5} \times \dfrac{4}{7}$
= _____

25) $\dfrac{7}{8} \times \dfrac{7}{14}$
= _____

Dividing Fractions

Write your answers as an improper fraction.

1) $\dfrac{2}{4} \div \dfrac{2}{3}$

= _____

2) $\dfrac{2}{4} \div \dfrac{1}{3}$

= _____

3) $\dfrac{1}{5} \div \dfrac{3}{4}$

= _____

4) $\dfrac{3}{4} \div \dfrac{2}{3}$

= _____

5) $\dfrac{3}{5} \div \dfrac{1}{2}$

= _____

6) $\dfrac{3}{10} \div \dfrac{3}{4}$

= _____

7) $\dfrac{6}{8} \div \dfrac{1}{7}$

= _____

8) $\dfrac{5}{14} \div \dfrac{1}{2}$

= _____

9) $\dfrac{1}{9} \div \dfrac{13}{18}$

= _____

10) $\dfrac{2}{3} \div \dfrac{14}{18}$

= _____

11) $\dfrac{9}{20} \div \dfrac{13}{18}$

= _____

12) $\dfrac{4}{6} \div \dfrac{10}{14}$

= _____

13) $\dfrac{15}{16} \div \dfrac{2}{6}$

= _____

14) $\dfrac{7}{16} \div \dfrac{2}{7}$

= _____

15) $\dfrac{2}{15} \div \dfrac{4}{5}$

= _____

16) $\dfrac{4}{5} \div \dfrac{10}{18}$

= _____

17) $\dfrac{4}{9} \div \dfrac{4}{7}$

= _____

18) $\dfrac{2}{3} \div \dfrac{2}{8}$

= _____

19) $\dfrac{10}{15} \div \dfrac{9}{18}$

= _____

20) $\dfrac{3}{6} \div \dfrac{5}{18}$

= _____

Dividing Mixed Numbers

1 $3\dfrac{5}{9} \div 3\dfrac{3}{7}$

= _____

2 $2\dfrac{1}{3} \div 2\dfrac{1}{4}$

= _____

3 $3\dfrac{5}{6} \div 4\dfrac{2}{5}$

= _____

4 $2\dfrac{1}{9} \div 3\dfrac{1}{7}$

= _____

5 $3\dfrac{1}{8} \div 2\dfrac{3}{5}$

= _____

6 $3\dfrac{1}{3} \div 3\dfrac{3}{4}$

= _____

7 $2\dfrac{1}{2} \div 2\dfrac{3}{7}$

= _____

8 $2\dfrac{7}{9} \div 3\dfrac{3}{7}$

= _____

9 $3\dfrac{1}{2} \div 4\dfrac{1}{3}$

= _____

10 $2\dfrac{5}{8} \div 2\dfrac{5}{7}$

= _____

11 $2\dfrac{1}{2} \div 2\dfrac{3}{7}$

= _____

12 $3\dfrac{5}{8} \div 3\dfrac{1}{6}$

= _____

13 $4\dfrac{2}{5} \div 2\dfrac{6}{7}$

= _____

14 $3\dfrac{2}{7} \div 4\dfrac{3}{5}$

= _____

15 $4\dfrac{1}{2} \div 3\dfrac{2}{3}$

= _____

Improper Fractions and Mixed Numbers

Convert the improper fraction to a mixed number or the mixed number to an improper fraction.

1) $\dfrac{26}{4}$ = _____

2) $\dfrac{52}{8}$ = _____

3) $\dfrac{22}{9}$ = _____

4) $\dfrac{14}{5}$ = _____

5) $\dfrac{36}{5}$ = _____

6) $\dfrac{37}{6}$ = _____

7) $\dfrac{53}{8}$ = _____

8) $\dfrac{40}{6}$ = _____

9) $\dfrac{22}{4}$ = _____

10) $\dfrac{29}{10}$ = _____

11) $4\dfrac{3}{4}$ = _____

12) $3\dfrac{2}{3}$ = _____

13) $4\dfrac{4}{5}$ = _____

14) $5\dfrac{4}{5}$ = _____

15) $6\dfrac{1}{2}$ = _____

16) $5\dfrac{5}{7}$ = _____

17) $7\dfrac{1}{4}$ = _____

18) $3\dfrac{1}{3}$ = _____

19) $7\dfrac{2}{5}$ = _____

20) $9\dfrac{1}{6}$ = _____

Comparing Decimals

Order the following decimal numbers from least to greatest value.

1 0.30, 0.6, 0.65

_____

2 0.60, 0.964, 0.546

_____

3 0.14, 0.23, 0.607

_____

4 0.76, 0.3, 0.63

_____

5 0.54, 0.860, 0.459

_____

6 0.02, 0.482, 0.11

_____

7 0.2, 0.40, 0.5

_____

8 0.881, 0.71, 0.725

_____

9 0.405, 0.50, 0.6

_____

10 0.7, 0.034, 0.9

_____

11 0.92, 0.67, 0.39, 0.91

_____

12 0.513, 0.689, 0.796, 0.514

_____

13 0.342, 0.421, 0.305, 0.422

_____

14 0.984, 0.989, 0.898, 0.98

_____

15 0.858, 0.908, 0.843, 0.921

_____

Adding Decimals

1
97.88
+ 19.37

= _____

2
69.62
+ 75.88

= _____

3
85.57
+ 55.92

= _____

4
69.54
+ 67.99

= _____

5
57.63
+ 43.68

= _____

6
66.32
+ 47.99

= _____

7
32.67
+ 18.87

= _____

8
21.75
+ 48.46

= _____

9
58.22
+ 95.89

= _____

10
88.77
+ 53.96

= _____

11
33.48
+ 68.88

= _____

12
47.949
+ 26.679

= _____

13
42.9947
+ 80.194

= _____

14
33.683
+ 33.357

= _____

15
86.416
+ 32.333

= _____

Subtracting Decimals from Whole Numbers

1
2
− 0.55

= _____

2
4
− 2.64

= _____

3
8
− 0.36

= _____

4
80
− 15.3

= _____

5
246
− 6.4

= _____

6
377
− 6.3

= _____

7
460
− 3.6

= _____

8
943
− 942.1754

= _____

9
9
− 6.302

= _____

10
5323
− 731.2

= _____

11
18
− 3.6949

= _____

12
38
− 18.2091

= _____

13
571
− 4.61

= _____

14
5
− 1.5011

= _____

15
70
− 6.662

= _____

Multiplying Decimals

1) 4.42
x 5.56
= _____

2) 6.23
x 4.37
= _____

3) 2.79
x 4.13
= _____

4) 2.68
x 4.99
= _____

5) 6.82
x 3.48
= _____

6) 7.19
x 7.14
= _____

7) 4.93
x 8.77
= _____

8) 9.22
x 7.72
= _____

9) 3.66
x 2.69
= _____

10) 2.54
x 3.38
= _____

11) 16.23
x 13.49
= _____

12) 80.79
x 10.25
= _____

13) 36.42
x 79.62
= _____

14) 29.52
x 29.98
= _____

15) 53.18
x 11.27
= _____

Dividing with Decimals

1   3.60
÷ 9

= _____

2   6.3
÷ 3

= _____

3   6.56
÷ 8

= _____

4   2.76
÷ 2

= _____

5   9.96
÷ 4

= _____

6   4.14
÷ 3

= _____

7   7.75
÷ 5

= _____

8   5.60
÷ 5

= _____

9   5.68
÷ 8

= _____

10   2.76
÷ 6

= _____

11   21.27
÷ 3

= _____

12   17.08
÷ 7

= _____

13   72.36
÷ 27

= _____

14   367.96
÷ 4

= _____

15   938.86
÷ 13

= _____

Mixed Operations with Decimals

Perform the indicated calculations to solve the problems presented.

1) 62.454
  + 36.748
  = _____

2) 77.165
  + 18.306
  = _____

3) 44.544
  - 22.354
  = _____

4) 7.18
  x 78
  = _____

5) 12.982
  + 15.484
  = _____

6) 39.638
  + 76.042
  = _____

7) 25.7
  x 4.5
  = _____

8) 9.51
  ÷ 6
  = _____

9) 48.726
  - 7.967
  = _____

10) 55.801
  - 49.884
  = _____

11) 4.67
  x 71
  = _____

12) 5.09
  ÷ 8
  = _____

13) 81.927
  + 62.768
  = _____

14) 49.7
  x 7.9
  = _____

15) 0.56
  ÷ 0.2
  = _____

16) 8.592
  - 5.394
  = _____

17) 84.882
  - 61.501
  = _____

18) 7.02
  ÷ 0.9
  = _____

19) 0.738
  x 0.87
  = _____

20) 507.535
  ÷ 853
  = _____

Decimal Word Problems: Working with Money

Use your knowledge of decimals to solve the following word problems involving money.

1 How much would it cost to buy two plain bagels for $0.79 each and one bagel sandwich for $3.50?

2 You start the day with $10. You buy milk at school for $0.65 and a baseball card from a friend after school for $4.75. How much do you have at the end of the day?

3 Your little brother needs help determining how much money he has in his piggy bank because he wants to buy your mom a birthday gift. He asks you to help him count the money. Together, you determine he has 5 quarters, 12 dimes, 17 nickels, and 84 pennies. How much does he have in total?

4 Jeff makes $8.25/hour watering plants. How much will he make for 7 hours of work?

5 A four-person relay team wins $15 in a race. They split the earnings equally. How much does each runner get?

6 Heather is buying pizza by the slice for her friends. How much does she need to spend to buy the following: 2 pieces of cheese for $1.75 each, 1 pepperoni for $2.00, 2 sausage and pepper for $2.10 each, 3 supreme for $2.75 each, and 1 portabella mushroom for $2.50.

7 Sales tax in Texas is 6.25%. If Colt wants to buy a hammock that is priced at $14.95, what should he expect to pay at the register with tax?

8 Lulu's mom gives her $20 to spend during summer sleepaway camp at the store. She gets the baseball cap for $9.75, 2 postcards for $0.65 each, a ballpoint pen for $0.18, and a keychain for $3.25. How much does her mom get back at the end of camp?

9 Three friends sell lemonade on the local bike path one afternoon. They sell 43 cups for $0.75 each. If they split the earnings equally, how much does each friend get?

10 Danica babysits 3.5 hours on Tuesday afternoon, 1.25 hours on Friday night, and 2.5 hours on Saturday morning. She gets paid $7.75/hour. How much did she earn in the week?

Decimal Word Problems: Working with Money

Use your knowledge of decimals to solve the following word problems involving money.

11 Tyler cleans his dad's car and finds loose change stuck in the seats, under the floor mats, in the doors, and around the console. He finds six pennies, four dimes, seven nickels, and five quarters. How much money did he find in all?

12 The local grocery store has just made a new policy. With the hopes of having customers be more eco-friendly and bring their own reusable canvas bags, the store now charges $0.05 for every plastic bag the customer uses for their groceries. If Mr. Robinson forgot his canvas bags and had his groceries bagged up in nine plastic bags, how much did he have to pay for the bags themselves?

13 Gloria buys milk at the cafeteria for lunch. The milk costs 35 cents. What are four possible combinations of coins she could use to give the cashier exact change for the milk?

14 Tyler buys a lollipop at the convenience store. He only had a dollar bill in his pocket. The cashier gives him two quarters and three dimes, and a nickel as change. How much was the lollipop?

15 Mike's family is having a yard sale. Mike is selling his old comic books for $0.10 each and his old picture books for $0.25 each. If he sold seven picture books and he earned a total of $2.65, how many comic books did he sell?

16 Walter's parents give him their pocket change to add to the collections basket at church on Sunday. His mom has two quarters, seven nickels, and three pennies. His dad has four quarters, one dime, and three nickels. How much money did Walter drop in the collection basket?

17 Popsicles at the town swimming pool are $0.75. How many quarters are needed to buy one if you only pay with quarters?

18 Evan's dad is trying to motivate him to read more books instead of play on his tablet. He decides to give Evan a nickel for every two pages Evan reads in his chapter book. If Evan reads 10 pages, how much money will he earn?

19 Lao and Peter set up a lemonade stand. At the end of the first hour, they sort the coins they've received. Can you help them determine how much money they've made so far? They have 12 quarters, 9 dimes, and 7 nickels.

20 Your little brother collects pennies. He asks you to help him understand how much money he has in terms of other coins. You help him count his pennies and find that he has 82 of them. If he traded these for other coins, what is the fewest number of coins he would get back and what would they be?

Converting Decimals to Fractions and Fractions to Decimals

Convert the following fractions into their equivalent decimal or decimal to its equivalent fraction. Be sure to reduce all fractions to lowest terms.

1 $\frac{1}{3}$ = _____

2 $\frac{6}{8}$ = _____

3 0.5 = _____

4 $\frac{1}{5}$ = _____

5 0.3 = _____

6 $\frac{6}{10}$ = _____

7 0.1 = _____

8 $\frac{4}{5}$ = _____

9 $\frac{1}{2}$ = _____

10 $\frac{5}{8}$ = _____

11 0.6 = _____

12 $\frac{2}{5}$ = _____

13 0.75 = _____

14 0.125 = _____

15 0.8 = _____

16 0.167 = _____

17 0.25 = _____

18 $\frac{1}{12}$ = _____

19 $\frac{7}{8}$ = _____

20 $\frac{2}{3}$ = _____

21 0.9 = _____

22 $\frac{6}{12}$ = _____

23 0.33 = _____

24 $\frac{3}{8}$ = _____

25 0.667 = _____

Converting Decimals to Fractions and Fractions to Decimals

Convert the following fractions into their equivalent decimal or decimal to its equivalent fraction. Be sure to reduce all fractions to lowest terms.

1 0.083

= _____

2 $\frac{3}{5}$

= _____

3 $\frac{3}{4}$

= _____

4 $\frac{4}{6}$

= _____

5 $\frac{5}{6}$

= _____

6 $\frac{3}{5}$

= _____

7 $\frac{1}{11}$

= _____

8 $\frac{7}{8}$

= _____

9 $\frac{4}{9}$

= _____

10 $\frac{17}{18}$

= _____

11 $\frac{10}{13}$

= _____

12 $\frac{12}{21}$

= _____

13 $\frac{1}{19}$

= _____

14 $\frac{9}{25}$

= _____

15 $\frac{2}{7}$

= _____

16 0.3

= _____

17 0.54

= _____

18 0.2

= _____

19 0.44

= _____

20 0.01

= _____

21 0.98

= _____

22 0.64

= _____

23 0.37

= _____

24 0.125

= _____

25 1.0

= _____

Complete the Pattern

Complete the following patterns.

1

2

3

4

5

6

7

8

9

10

Numerical Series

Complete the numerical series by filling in the blanks.

1) 15, 10, 20, 15, 25, _____, _____, _____

2) 1, 3, 5, 7, _____, _____, _____

3) 4, 8, 12, _____, 20, _____, 28, _____

4) 2, 4, 8, 16, 32, _____, _____, _____

5) 85, 79, 73, 67, 61, _____, _____, _____

6) 3, 14, 25, 36, 47, _____, _____, _____

7) 99, 90, 81, 72, _____, _____, _____

8) 0, 1, 1, 2, 3, 5, 8, 13, _____, _____, _____

9) 46, 43, 40, 37, _____, _____, _____

10) 2, 6, 18, 54, _____, _____, _____

11) 9, 12, 7, 10, 5, 8, _____, 6, _____, 4, _____

12) 56, 49, _____, 35, 28, _____, _____

13) 6, 7, 9, 12, 16, _____, 27, _____, 42, _____

14) 1, 4, 9, 16, _____, 36, _____, 64, _____

15) 30, 22, 14, 6, -2, _____, _____, _____

16) 2, 4, 14, 16, 26, 28, _____, 40, _____, _____

17) 21, 27, 22, 28, 23, 29, _____, _____, _____, 31

18) 160, 80, 40, _____, _____, 5, _____, 1.75

19) 1, 2, 4, 8, 10, 20, _____, _____, 46, _____

20) 2, 3, 6, 11, 18, _____, _____, _____, 66

21) 5, 15, 18, 54, 57, 171, _____, _____, _____ 1575

22) 4, 8, 3, 6, 1, 2, -3, _____, _____, _____

23) 1, 2, -4, 8, -16, _____, _____, _____

24) 14, 21, 16, 23, 18, 25, _____, _____, _____, _____

25) 2, 4, 6, 12, 14, 28, _____, _____, _____, 124

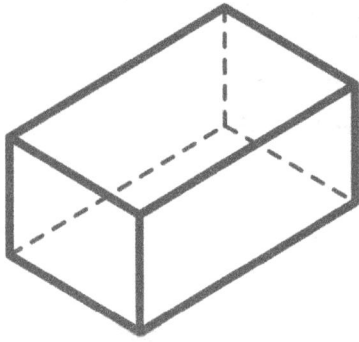

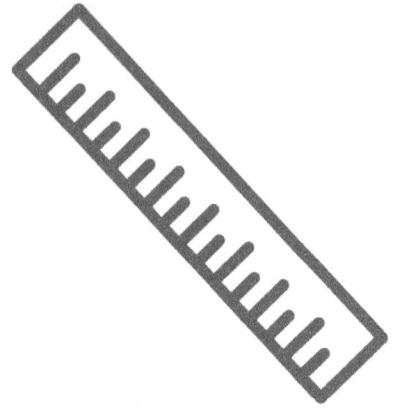

Measurement and Data

Reading Measurements in Two Units

For each of the rulers below, read the length of the line to the nearest half-inch and nearest half-centimeter. Record both measurements. Answer the questions at the end.

1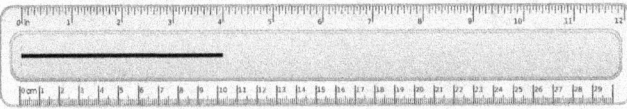

Length: _____ inches,

_____ centimeters

2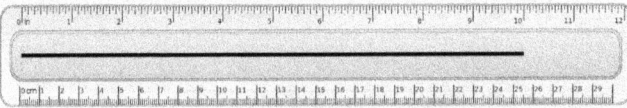

Length: _____ inches,

_____ centimeters

3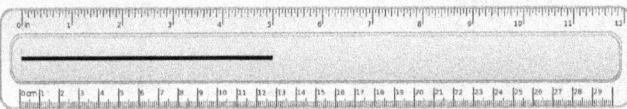

Length: _____ inches,

_____ centimeters

4

Length: _____ inches,

_____ centimeters

5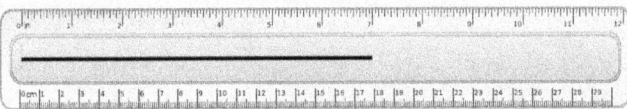

Length: _____ inches,

_____ centimeters

6

Length: _____ inches,

_____ centimeters

7

Length: _____ inches,

_____ centimeters

8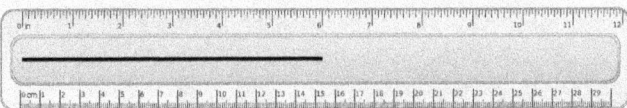

Length: _____ inches,

_____ centimeters

Reading Measurements in Two Units

For each of the rulers below, read the length of the line to the nearest half-inch and nearest half-centimeter. Record both measurements. Answer the questions at the end.

9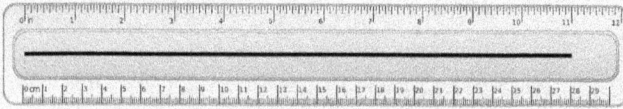

Length: _____ inches,

_____ centimeters

10

Length: _____ inches,

_____ centimeters

11

Length: _____ inches,

_____ centimeters

12

Length: _____ inches,

_____ centimeters

13 Which is larger?

3 inches   6 centimeters

14 Which is larger?

4 inches   11 centimeters

15 Which of the following is true?

a. There are about 2.5 inches in 1 centimeter

b. There are about 2.5 centimeters in 1 inch

c. There are about 10 inches in 1 centimeter

d. There are about 10 centimeters in 1 inch

Getting Familiar With Solid Figures

For each solid, count the number of faces, vertices, and edges and write the name of the figure on the provided line.

1

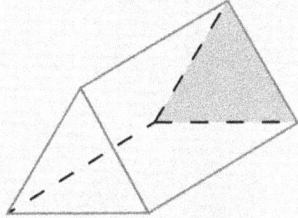

Name: _____

Number of Faces: _____

Number of Vertices: _____

Number of Edges: _____

2

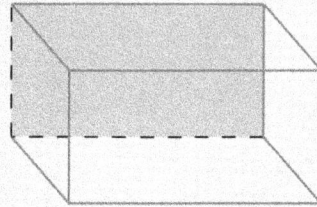

Name: _____

Number of Faces: _____

Number of Vertices: _____

Number of Edges: _____

3

Name: _____

Number of Faces: _____

Number of Vertices: _____

Number of Edges: _____

4

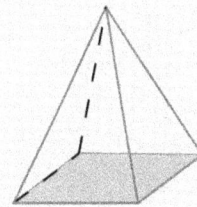

Name: _____

Number of Faces: _____

Number of Vertices: _____

Number of Edges: _____

5

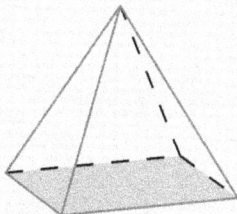

Name: _____

Number of Faces: _____

Number of Vertices: _____

Number of Edges: _____

6

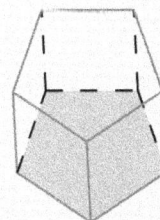

Name: _____

Number of Faces: _____

Number of Vertices: _____

Number of Edges: _____

Getting Familiar With Solid Figures

For each solid, count the number of faces, vertices, and edges and write the name of the figure on the provided line.

7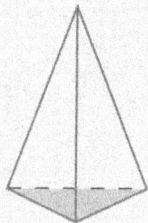

Name: _____

Number of Faces: _____

Number of Vertices: _____

Number of Edges: _____

8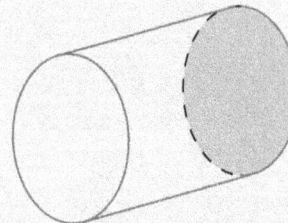

Name: _____

Number of Faces: _____

Number of Vertices: _____

Number of Edges: _____

9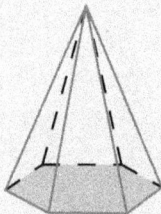

Name: _____

Number of Faces: _____

Number of Vertices: _____

Number of Edges: _____

10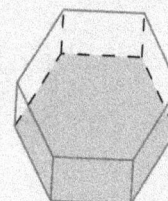

Name: _____

Number of Faces: _____

Number of Vertices: _____

Number of Edges: _____

Calculating the Volume of Rectangular Prisms

Determine the volume (in cubic units) of the following rectangular prisms

1

6
6
5

_____
_____
_____

2

6
2
2

_____
_____
_____

3

3
3
1

_____
_____
_____

4

5
4
2

_____
_____
_____

5

5
8
4

_____
_____
_____

6

6
4
2

_____
_____
_____

7

1
20
7

_____
_____
_____

8

2
10
2

_____
_____
_____

9

2
8
9

_____
_____
_____

10

8
7
5

_____
_____
_____

Calculating the Volume of Rectangular Prisms

Determine the volume (in cubic units) of the following rectangular prisms

11 The side length of the cube below is 5 centimeters. What is the volume?

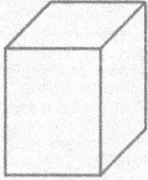

_____
_____
_____
_____

12 The side length of the cube below is 8 centimeters. What is the volume?

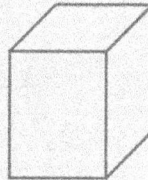

_____
_____
_____
_____

13 The volume of the cube below is 8 cubic centimeters. What is the length of one side?

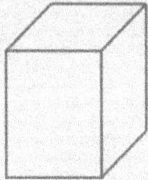

_____
_____
_____

14 The volume of the cube below is 343 cubic centimeters. What is the length of one side?

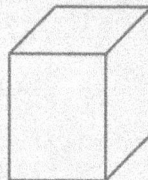

_____
_____
_____
_____

15 The volume of the rectangular prism below is 24 cubic centimeters. What the length of the missing side?

3
1

_____
_____
_____
_____

16 The volume of the rectangular prism below is 360 cubic centimeters. What is the length of the missing side?

8
5

_____
_____
_____
_____

17 The volume of the rectangular prism below is 42 cubic centimeters. What is the length of the missing side?

3
2

_____
_____
_____
_____

18 The volume of the rectangular prism below is 192 cubic centimeters. What is the length of the missing side?

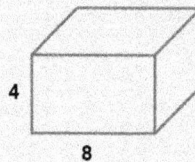

4
8

_____
_____
_____
_____

19 The volume of the rectangular prism below is 450 cubic centimeters. What is the length of the missing side?

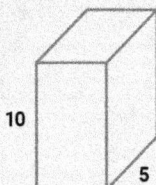

10
5

_____
_____
_____
_____

20 The volume of the cube below is 64 cubic centimeters. What is the length of one side?

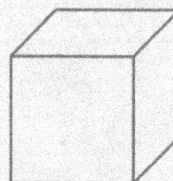

_____
_____
_____
_____

Calculating the Volume of Real-World Solids

Find the volumes of the following solids. Use 3.14 for π where necessary.

1 A small cube-shaped tissue box that has a side length of 12 cm.

2 A ring box with a side length of 18 mm.

3 A cereal box that is 3 inches wide, 7 inches wide, and 11 inches tall.

4 A kid's soccer ball with a diameter of 8 inches.

5 A rectangular swimming pool that is 25 yards long, 12 yards wide, and 2 yards deep.

6 The Great Pyramid with a height of 146 m and a side of the square base of 230 m.

7 A snow cone with a length (height) of 5 inches and a radius of 2 inches.

8 A chocolate box with a length of 12 inches, a width of 8 inches, and a depth of 3 inches.

9 A sand castle in the shape of a rectangular pyramid with a base side lengths of 5 inches and 9 inches and a height of 10 inches.

10 A cylindrical water tank that is 80 meters high and has a radius of 12 meters.

11 A shipping box that is 15 inches by 7 inches by 8 inches.

12 A fish tank that is 24 inches by 18 inches by 16 inches.

13 A play teepee in the shape of a cone that is 6 feet high and has a diameter of 8 feet.

14 A farmer's grain storage container (cylindrical) that is 8 meters high and has a diameter of 4 meters.

15 A jewelry box with a length of 10 inches, a depth of 5 inches, and a width of 7 inches.

16 A can of soup that is 4 inches tall and has a radius of 2 inches.

17 The sphere of a gumball machine that has a radius of 5 inches.

18 A little bouncy rubber ball with a diameter of 6 cm.

19 A traffic cone with a height of 36 inches and a diameter of 12 inches.

20 A large gourmet spherical lollipop with a radius of 1 inch.

Weight Word Problems

1 Kimberly's baby sister drinks 4 ounces of milk every 3 hours. How much milk, in ounces, does she drink in 24 hours?

_____

2 A crate of oranges weighs 14.4 kilograms. If it holds 40 oranges, what is the average weight of an orange?

_____

3 Tom weighs 45.2 kg. If he gains 1785 grams, how much does he weight now?

_____

4 A suitcase can hold up to 50 pounds for a certain airline. If Jack's suitcase was 43 pounds 6 ounces on the way to San Diego, what is the maximum weight he can add to his suitcase in souvenirs before his trip home?

_____

5 A bookshelf weighs 18 pounds 7 ounces. An end table weighs 4 pounds 9 ounces. How much more does the bookshelf weigh than three end tables?

_____

6 A candle maker adds 30 grams of wax to each little candle mold. If he melts 6.5 kg of wax, how many full candle molds can he fill?

_____

7 Sheila is donating canned goods to the food pantry. One box of cans weighs 20 pounds 14 ounces. Another box weighs 14 pounds 5 ounces, and the third box weighs 15 pounds 8 ounces. What is the total weight of food she is donating?

_____

8 At an apple farm, large bags of apples weigh 15 pounds 5 ounces and small bags weigh 9 pounds 9 ounces. What is the total weight of Adam's purchase if he gets 3 large bags and 3 small bags?

_____

Weight Word Problems

9 Darrel is weighing insects in science class using a microscale. The total weight of the specimens is 894 grams. Two insects weigh 48.6 grams each, one insect weighs 54.3 grams, three beetles weigh 84.8 grams, a butterfly weighs 103.9 grams, and two caterpillars each weigh 98.4 grams. What is the weight of the final praying mantis?

_____

10 Malcom used 9 oz. of butter In a batch of 12 cupcakes. He used 2 pounds of butter in a batch of 46 cookies. Which has more butter per unit—a cupcake or a cookie?

_____

11 How many 450-gram bowls of cereal can a cafeteria serve from each economy-size 13-kg box of cereal?

_____

12 The weight capacity of a certain truck is 2000 pounds. The truck currently has 15 refrigerators that each weigh 96 pounds. How many more refrigerators can the truck hold?

_____

13 Gregory's puppy weighed 7 pounds 2 ounces last week. This week, he weighed 9 pounds 1 ounce. If the puppy gains the same amount of weight next week, how much will he weigh?

_____

14 Each kitten in a cat's litter weighs 9 ounces. If the mother cat has 12 kittens, what is the total weight of the litter in pounds?

_____

15 Jim's dad is making hamburgers for a cookout. If he has five pounds of meat and he needs to make 15 burgers of equal size, what is the weight of each burger in ounces?

_____

Length Word Problems

1 Sandra has three strips of red ribbon that are each 35 inches and two strips of blue ribbon, each of which is 18 inches. How much ribbon does she have in total?

_____

_____

2 The flagpole at school is 30 feet 5 inches tall. The parking lot lamps are 26 feet 9 inches tall. How much taller is the flagpole?

_____

_____

3 Tommy is a long jumper on his track team. He wants to jump 18 feet. His first practice jump was nine feet nine inches. How much further does he need to jump as he improves to reach his goal?

_____

_____

4 Jeffrey is lining up leaves he collected in his backyard. He places them end to end along a yardstick (which is 36 inches long) so he can see the total length of leaves he has collected. He has 1 oak leaf that is 4 inches, 2 maple leaves that are each 3 inches, a beech tree leaf that is five inches, and two Japanese maple leaves that are each 4 inches long. What is the total length of his leaves.

_____

_____

5 In the problem above, what length of the yardstick is left without Jeffrey's leaves covering it?

_____

_____

6 Chandra has been growing out her hair to donate it to a charity that makes wigs for kids under-going cancer treatments. When she goes to the barber shop, they measure her hair to be 24 inches long. If her hair is 1/3 as long after the cut, what length of hair did she donate?

_____

_____

7 Tommy and his little brother, Bert, are comparing the lengths of their feet. They trace their bare feet and then measure the lengths of the tracings with a ruler. Tommy's feet are 9 inches and Bert's feet are two-thirds that length. How long are Bert's feet?

_____

_____

Length Word Problems

8 Gwen's dad is extending a fence around the garden because they are going to make the plot bigger. Currently, the fence is 12 feet along the length and 5 feet along the width and he wants double each dimension. What will be the new length of the entire fence?

_____

_____

9 Nora runs 5.5 miles on Saturday, 6.75 miles on Sunday, and 3.75 miles on Monday. How far did she run in total over the three days?

_____

_____

10 Gwen has two Siamese cats named Miso and Tamari. Miso's tail is 10 inches and Tamari's tail is ¾ as long. How much longer in inches is Miso's tail?

_____

_____

11 A yardstick is 36 inches long and a standard ruler is 12 inches long. If a fifth grade classroom has 3 yardsticks and 25 rulers, what is the total length end to end?

_____

_____

12 Dad's commute to work is 28 miles each way and Mom has to drive 19 miles each way to her job. In a five day work week, round trip, how much further does Dad drive?

_____

_____

13 Benji bikes 4 miles to the park from his house. From there, he bikes 10.5 miles with his friend, Joyce, to get ice cream. They eat their ice cream then bike to Joyce's house to watch a movie. The distance from the ice cream place to Joyce's house was half the distance that the park was from ice cream shop. How far did Benji bike in total?

_____

_____

14 Jessica's Nana is knitting a baby blanket. She wants it to be 48 inches long. She has knitted 1/4 of it so far. How much does she have left in inches?

_____

_____

Length Word Problems

15 Ms. Landin is laminating artwork that her students made. Jenny's picture is on paper that is 11 inches long. Koi's drawing is on paper that is 15 inches long, and Peter's little drawing is half the length of Koi's. What length of laminating plastic will she need if the drawings are lined up right next to each other?

_____

_____

16 Dennis is having the border of his bedroom along the top of the wall stenciled with sailboats. The total length around the four walls of his room is 44 feet. Each sailboat is 4 inches wide. How many sailboats will go around the room if there are no gaps between them?

_____

_____

17 Fiona and Freddie are competing against one another to see who can throw a Frisbee farther. They each get three throws and they are comparing the total distance of the three throws. Fiona throws hers 57 feet, 49 feet, and 54 feet. Freddie throws his 63 feet, 69 feet, and only 29 feet. How far did each person throw?

_____

_____

18 In the above scenario, who won and by how much?

_____

_____

19 Jasim's family strings popcorn and cranberries to decorate their Christmas tree. Jasmin's string is 6 feet. His mom's string is one-and-a-half times as long. His brother's string is 9 feet, and his dad's string is twice as long as that. What length do they have in all?

_____

_____

20 A football team makes a successful pass and gains 27 yards. On the next play, they gain 1/3 as many more yards. How much have they gained in the two plays together?

_____

_____

Estimating Measurements

Chandra has lost her measuring tools so she needs to estimate the lengths of various objects. Help her by circling the most likely measurement for her object out of the options provided.

1 A man's sneaker

a. 1 gram　　b. 18 kg
c. 18 ounce　d. 18 pound

2 An elephant

a. 100 pounds　b. 1 ton
c. 1000 tons　d. 1000 grams

3 A bicycle

a. 18 pounds　b. 18 ounces
c. 18 grams　　d. 18 kilograms

4 A brick

a. 1 pound　　b. 5 pounds
c. 50 grams　d. 50 ounces

5 A slice of bread

a. 1 gram　　b. 1 kilogram
c. 1 ounce　d. 1 pound

6 A newborn baby

a. 1 pound　　b. 1 kg
c. 8 pounds　d. 8 kg

7 A soda at the movie theater

a. 8 mL　　b. 800 mL
c. 8 liters　d. 80 cups

8 A can of paint for a wall

a. 5 mL　　b. 50 mL
c. 5 liters　d. 50 liters

9 A juice box

a. 2 gallons　b. 2 liters
c. 200 mL　　d. 20 mL

10 A small fish bowl for one goldfish

a. 1 gallon　　b. 10 gallons
c. 1 liter　　d. 10 liters

11 A full size candy bar

a. 8 inches　　b. 8 centimeters
c. 2 inches　　d. 2 centimeters

12 A new crayon out of the box

a. 4 centimeters　b. 1 inch
c. 10 centimeters　d. 1 foot

13 An average-sized banana

a. 1 foot　　b. 8 inches
c. 1 yard　　d. 8 centimeters

14 A football field

a. 100 feet　　b. 100 yards
c. 100 inches　d. 100 centimeters

15 A piece of regular notebook paper or computer paper the long way

a. 2 feet　　b. 3 feet
c. 1 inch　　d. 11 inches

16 The length of a fingernail

a. 5 inches　　b. 5 centimeters
c. 8 inches　　d. 1 centimeter

17 The kite string when the kite is soaring up in the sky

a. 5 feet　　b. 50 inches
c. 20 yards　d. 20 centimeters

18 A refrigerator

a. 6 inches　　b. 26 inches
c. 16 feet　　d. 6 feet

19 The height of a soda can

a. 5 inches　　b. 5 centimeters
c. 15 inches　　d. 1 foot

20 The height of a basketball hoop

a. 25 inches　　b. 25 feet
c. 10 inches　　d. 10 feet

()

x^2

Algebra

×

÷

±

Algebraic Word Problems

Solve each of the following word problems.

1 Carly purchased 84 bulbs for her flower garden. Tulips came in trays containing six bulbs and daffodils came in trays containing 8 bulbs. Carly bought an equal number of tulip and daffodil trays. How many of each type of flower bulb were purchased?

_____

2 Apples cost $2 each, while bananas cost $3 each. Maria purchased 10 fruits in total and spent $22. How many apples did she buy?

_____

3 Jessica buys 10 cans of paint. Red paint costs $1 per can and blue paint costs $2 per can. In total, she spends $16. How many red cans did she buy?

_____

4 Cameron sold half his yo-yo collection then bought six more. He now has 16 yo-yos. How many did he begin with?

_____

5 The sum of three consecutive pages in a book is 144. What is the lowest page number in the set?

_____

6 Jane got 9 tickets from winning a game at the arcade and adds them to her pile. She then spends half of her tickets on a mini football, leaving just 26 left. How many tickets did she have before buying the football?

_____

7 Noah had $165 to spend on 8 identical posters to give as favors at his birthday party. After buying the posters, he had 29 dollars left. How much was each poster?

_____

8 Three-hundred-thirty-one third grade students are taking a field trip to the local science museum. They are filling 7 buses and then 9 students are riding in an accessible van. How many students fit on each bus?

_____

9 After seeing a movie, Dennis, Jan, and Tina decide to split the total cost of the tickets and the snacks. If they each had to pay $13, and the snacks cost $12 total, how much was each ticket?

_____

Algebraic Word Problems

Solve each of the following word problems.

10 Tiara is selling cups of lemonade for $0.75. If each pitcher makes 8 cups of lemonade and the cost of ingredients is $1.12 per pitcher, how many cups does she need to sell to make $20 profit?

_____

11 Peter and his sister, Lisa, are collecting bottles and cans for redemption and plan to donate the proceeds to the local animal shelter. The redemption center gives $0.05 for each aluminum can, but glass bottles earn $0.10. If they bring their bottles and cans to the redemption center and get $29.00 total after redeeming ¾ as many glass bottles as cans, how many bottles did they redeem?

_____

12 Mei is baking brownies and cookies for a bake sale. The recipe for brownies makes 16 brownies and the recipe for cookies makes 24 cookies. How many batches of cookies does she need to make if she is to bring 184 treats and she makes one more batch of cookies than brownies?

_____

13 Dwayne's baseball team has 44 games on the schedule this season. If ¼ are home games, how many games are away?

_____

14 Juanita is counting ladybugs and spiders that she finds in her backyard. Altogether, she counts 198 legs. If she sees twice as many spiders as ladybugs, how many ladybugs were there?

_____

15 Mr. Read's seventh grade class is studying geography. If he is creating equal-sized groups to present on each of the seven continents and each group has four students, how many students are in the class?

_____

16 Shankar is training for a 10k road race. If he runs 5 miles in 42:30, then how long will it take to run 7 miles if he maintains the same pace?

_____

17 Becca practices piano 7 days a week. Some days, she plays 30 minutes, and some days, she plays 50 minutes. If she played 5 hours and 10 minutes in the week, how many days did she play 50 minutes?

_____

18 Sam's mom works at the local diner. She makes $8 per hour as a base wage plus tips. Last week, she earned $396. If tips made up 1/3 of her pay, how many hours did she work?

_____

Algebraic Word Problems

Solve each of the following word problems.

19 Valencia babysits a family with 3 kids. She makes $12 per hour. However, when one of the kids had a friend over, she makes an extra $3 per hour. Last month, she babysat 6 times. If one of those times there was an additional child, how much did she make if each occasion was 4 hours (including the one with a friend), except one occasion, which was 6 hours?

_____

20 The soccer team is selling donuts to raise money to buy new uniforms. For every box of donuts that they sell, the team receives $3 towards their new uniforms. There are 15 people on the team. How many boxes does each player need to sell in order to raise $270 for their new uniforms?

_____

21 At the store, Jan spends $90 on apples and oranges. Apples cost $1 each and oranges cost $2 each. If Jan buys the same number of apples as oranges, how many oranges did she buy?

_____

22 Kristen purchases $100 worth of CDs and DVDs. The CDs cost $10 each and the DVDs cost $15. If she bought four DVDs, how many CDs did she buy?

_____

23 In Jim's school, there are 3 girls for every 2 boys. There are 650 students in total. How many students are girls?

_____

24 Kimberley earns $10 an hour babysitting, and after 10 p.m., she earns $12 an hour, with the amount paid being rounded to the nearest hour accordingly. On her last job, she worked from 5:30 p.m. to 11 p.m. In total, how much did Kimberley earn for that job?

_____

25 Store brand coffee beans cost $1.23 per pound. A local coffee bean roaster charges $1.98 per $1\frac{1}{2}$ pounds. How much more would 5 pounds from the local roaster cost than 5 pounds of the store brand?

_____

26 Paint Inc. charges $2000 for painting the first 1,800 feet of trim on a house and $1.00 per foot for each foot after. How much would it cost to paint a house with 3125 feet of trim?

_____

27 Sam is twice as old as his sister, Lisa. Their oldest brother, Ray, will be 25 in three years. If Lisa is 13 years younger than Ray, how old is Sam?

_____

Order of Operations

Evaluate the following problems using the order of operations

1 $(10 - 3) \times (9 - 6) + 7^2$

2 $(15 - 7) \times (12 - 6) + 6^2$

3 $(10 - 3)^2 + (12 - 15 \div 5)$

4 $2 \times (6 \times 3 - 8^2) + 22$

5 $(9 + 56 - 5) \div 2 + 3^2$

6 $(11 + 53 - 4^2) \div (9 + 7)$

7 $(14 + 19 - 3^2) \div (12 \div 2)$

8 $(3^2 - 4)^2 + (16 + 20 \div 10)$

9 $3(8 \div 2^2)^3 + (5 \times 8)$

10 $(2 \times 6^2 + 8) \div (8 - 20 \div 5)^2$

11 $4 + (3 \times 2)^2 \div 4$

12 $2 \times (6 + 3) \div (2 + 1)^2$

13 $2^2 \times (3 - 1) \div 2 + 3$

14 $(12 + 3) \times (8 - 2) - 5^2$

15 $(19 - 8) \times (13 - 3) + 2^2$

16 $(2 + 4)^2 + (9 + 12 \div 4)$

17 $3 \times (13 \times 3 + 8^2) - 12$

18 $[(4 + 3)^2 + 1] + 2^3 - 5$

19 $[6^2 + (20 \div 5 + 4^2)] \div 7$

20 $(15 \div 5)^2 - [(12 + 2) + 3^2]$

Translating Algebraic Expressions

Translate the following verbal statements into mathematical expressions

1 2 more than x

_____

2 t minus 5

_____

3 Six times the sum of 12 and m

_____

4 The quotient of five and a

_____

5 Take away 4 from b

_____

6 The product of h and 11 is subtracted from two-fifths of j

_____

7 Three-fourths of d is subtracted from 9

_____

8 Four divided by the product of nine and s

_____

9 6 more than two-thirds of p

_____

10 Add 4 to 7 times g

_____

11 5 less than the sum of one-half of c and two times d

_____

12 Subtract one-third from 4 times b

_____

13 One-sixth of the sum of 8 and z minus the product of 2 and x

_____

Translating Algebraic Expressions

Translate the following verbal statements into mathematical expressions

14 8 is divided by the sum of three and x squared

_____

15 The sum of three-fourths of n, one-third of y, and 7

_____

16 5 less than 9 times k

_____

17 12 is added to the product of b and 4

_____

18 a squared plus the product of 7 and d

_____

19 Subtract 9 from 3 times p and divide the result by 2

_____

20 2 times the sum of 8 and r

_____

21 Three-fifths z times the difference of y and 1

_____

22 The difference of x times 4 and 14 times y squared

_____

23 2 less than the sum of one-fifth of g and two-thirds of m

_____

24 One-half of g is added to the quotient of 10 and a

_____

25 15 more than the product of d and 2 less than v

_____

Evaluating Algebraic Expressions

Evaluate the following algebraic expressions for the given values.

1 What is the value of $7b - 2a$ when $a = 8$, $b = 7$?

2 What is the value of $-3 - 9 - 6c + 7d$ when $c = 4$, $d = 8$?

3 What is the value of $6x + 7y$ when $x = 5$, $y = 3$?

4 What is the value of $-2(8m - 6n)$ when $m = 3$, $n = 6$?

5 What is the value of $x^2 - 2xy + 2y^2$ when $x = 2$, $y = 3$?

6 What is the value of $8n + 5n^3 + 16n^2$ when $n = 4$?

7 What is the value of $(15 - 8t^2)-(5g^3 - 9 + 6g^2)+(3 + 7t)$ when $t = -2$, $g = 7$?

8 What is the value of $(2a^2 + 6a^4 - 4a)(3b^3 + 8b^2 + b)$ when $a = 3$, $b = -6$?

9 What is the value of $9k^2 - 6l^2 + 8k$ when $k = 12$, $l = -8$?

10 What is the value of $(8b^2 + 3)+(58 - 2b^3)-(6b^3 + 4b)$ when $b = 2$?

11 What is the value of $(8c^3 - 6c^2 + 4)+(3d^3 + 7c^2)$ when $c = -4$, $d = 3$?

12 What is the value of $y(9 - 7x^4 + 2x)$ when $x = -3$, $y = 7$?

13 What is the value of $(8 + 4c^2)-(2d^3 - 3d^2)$ when $c = 15$, $d = 4$?

14 What is the value of $(2m^2 + 3)(4n^2 - 7n)$ when $m = 1$, $n = -1$?

Solving Equations

Solve each equation for the unknown variable.

1 $-5x = -40$

_____

2 $2 + j = -8$

_____

3 $-7 + a = -10$

_____

4 $-8h + 6h = 22$

_____

5 $12 = m - 2$

_____

6 $11 = c - 3$

_____

7 $-12 = 2y$

_____

8 $7f = 56$

_____

9 $-34 = 6.8c$

_____

10 $6g = 36$

_____

11 $\frac{z}{4} = 7$

_____

12 $-5.1d = -35.7$

_____

13 $\frac{b}{3} = -4$

_____

14 $2.5t = 10$

_____

15 $46.4 = -5.8a$

_____

16 $\frac{j}{5} = 4.2$

_____

17 $31.8 = 5.3x$

_____

18 $-16 = -2b - 4 + 5b$

_____

19 $9 = -23u - 22$

_____

20 $23 = 13d + 2$

_____

Geometry

Classifying Angles

Classify each of the following angles as acute, right, or obtuse based on eyeballing their measure.

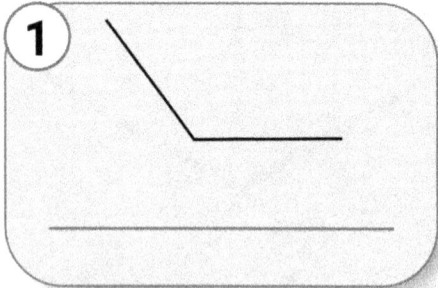

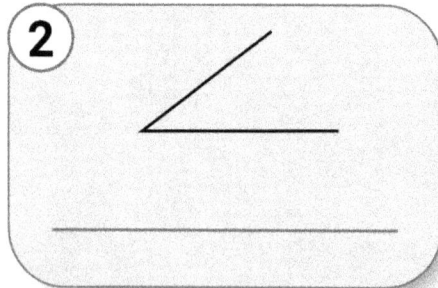

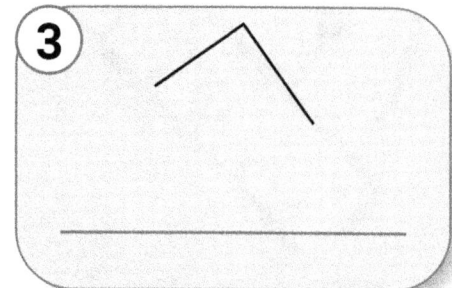

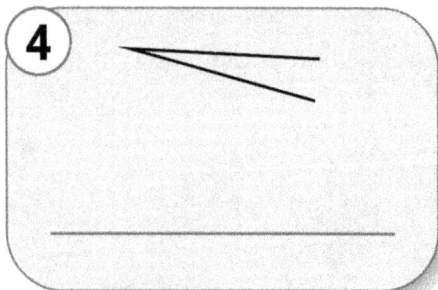

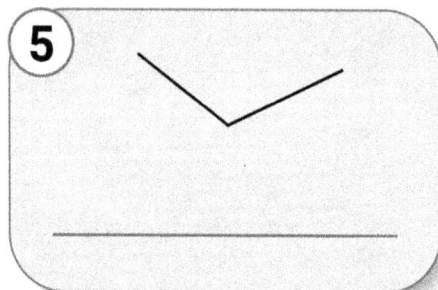

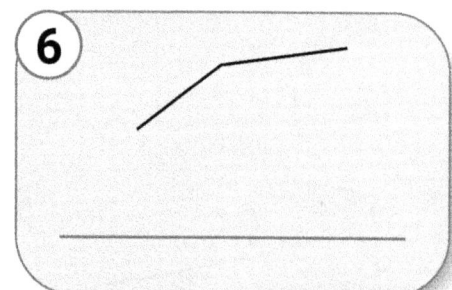

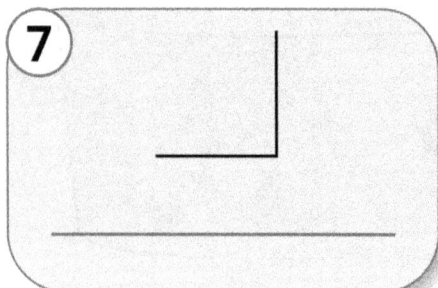

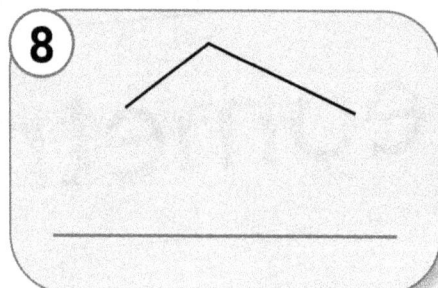

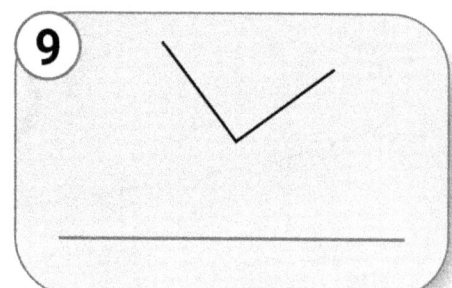

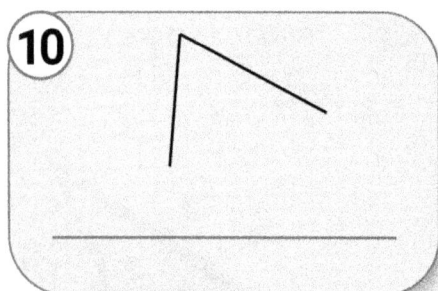

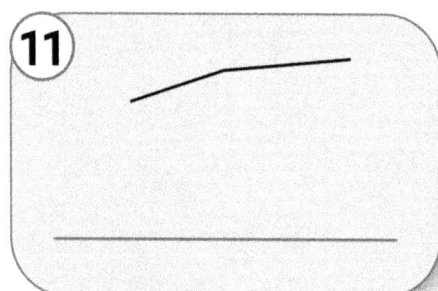

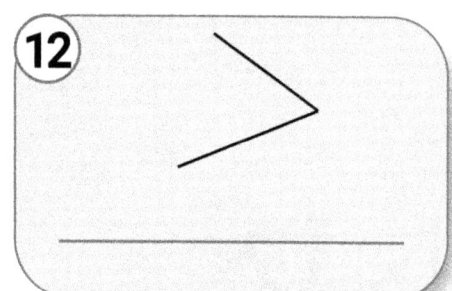

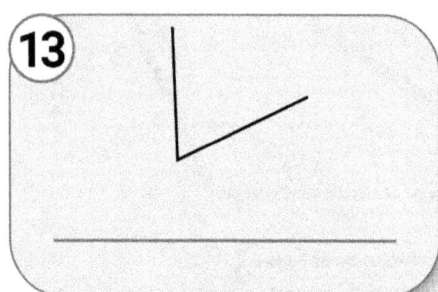

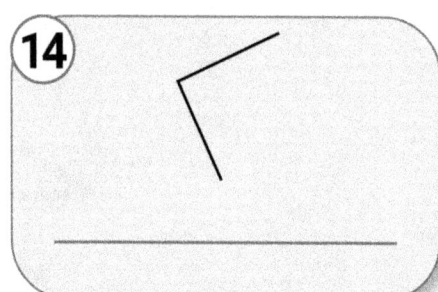

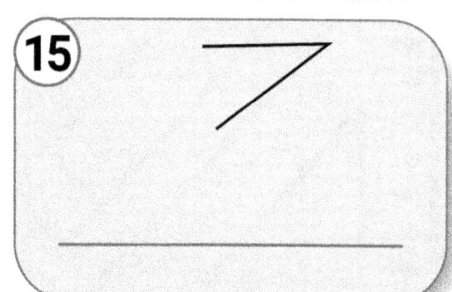

1

_____

2

_____

3

_____

4

_____

5

_____

6

_____

7

_____

8

_____

9

_____

10

_____

11

_____

12

_____

13

_____

14

_____

15

_____

Determining the Measurement of Missing Angles

Find the missing measures of the indicated angles.

1 What is the measure of angle ABD if angle ABC = 55°?

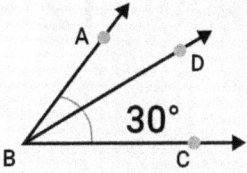

2 What is the measure of angle GHJ if angle GHK = 120°?

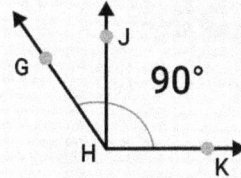

90°

3 What is the measure of angle ABC if angle ABD is 97°?

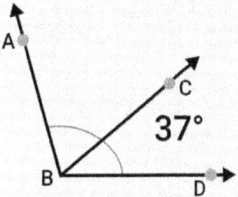

37°

4 What is the measure of angle RSU if angle TSU measures 65°?

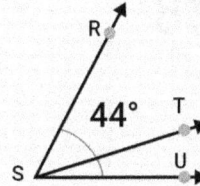

44°

5 Angle ACD measures 116°. It is divided into two smaller angles: Angle ABC measures 48°. Angle BCD, the measure of the other small angle, is unknown. What is the measure of angle BCD if angle ABC plus angle BCD equals angle ACD?

6 What is the measure of angle VYZ if angle XYZ = 23°, and angle XYV = 11°?

7 Angle LMO = 84°. It is divided into two smaller angles: Angle LMN measures 35°. Angle MNO, the measure of the other small angle, is unknown. What is the measure of angle MNO if angle LMN plus MNO equals angle?

8 What is the measure of angle CEF if angle DEF = 91°, and angle CED = 48°?

9 What is the measure of angle IJK if angle HIJ = 12°, and angle HIK = 27°?

10 What is the measure of angle GHJ if angle GHK = 125° and angle JHK is 92°?

Complementary and Supplementary Angles

Find the Complementary Angle
For each angle given, determine the measure of the complementary angle.

1. 10° _____

2. 55° _____

3. 30° _____

4. 80° _____

5. 12° _____

6. 27° _____

7. 38° _____

8. 71° _____

9. 63° _____

10. 56° _____

Find the Supplementary Angle
For each angle given, determine the measure of the supplementary angle.

1. 45° _____

2. 154° _____

3. 39° _____

4. 167° _____

5. 19° _____

6. 90° _____

7. 136° _____

8. 73° _____

9. 83° _____

10. 116° _____

11. 70° _____

12. 128° _____

13. 87° _____

14. 179° _____

15. 119° _____

Finding Missing Angles in Quadrilaterals

Find the measurement of the missing angle, x, in each the following quadrilaterals.
The sum of the interior angles on any quadrilateral is 360 degrees.

1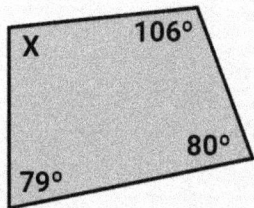

x = _____

2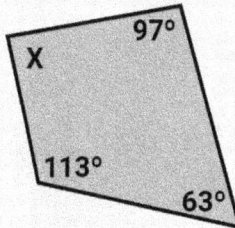

x = _____

3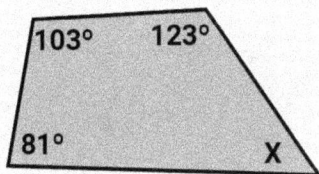

x = _____

4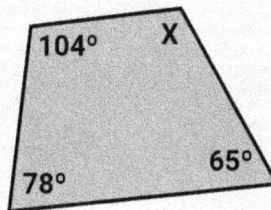

x = _____

5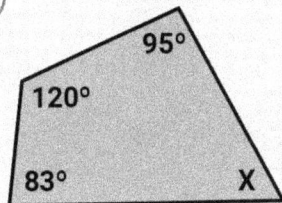

x = _____

6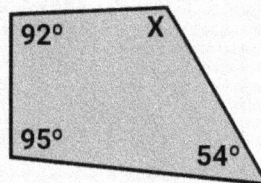

x = _____

7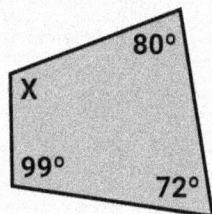

x = _____

8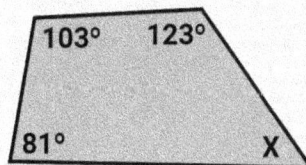

x = _____

9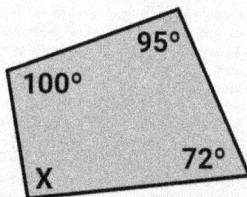

x = _____

10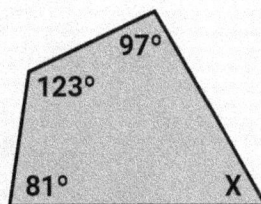

x = _____

Finding Missing Angles in Quadrilaterals

Find the measurement of the missing angle, x, in each the following quadrilaterals.
The sum of the interior angles on any quadrilateral is 360 degrees.

11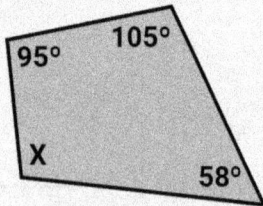

95° 105°

X

58°

x = _____

12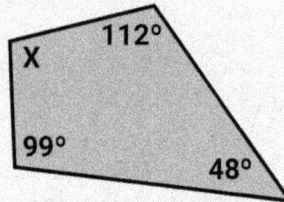

X 112°

99°

48°

x = _____

13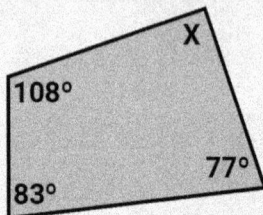

X

108°

77°

83°

x = _____

14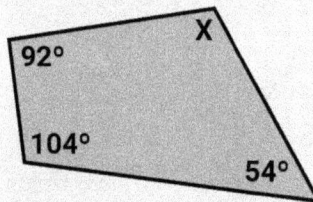

92° X

104°

54°

x = _____

15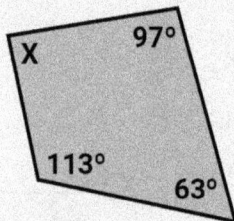

X 97°

113°

63°

x = _____

16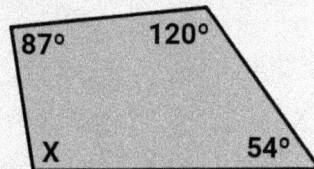

87° 120°

X

54°

x = _____

17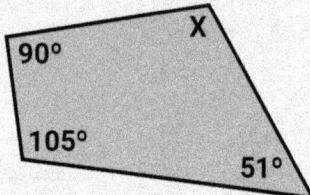

X

90°

105°

51°

x = _____

18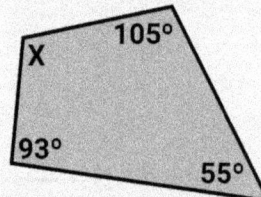

X 105°

93°

55°

x = _____

19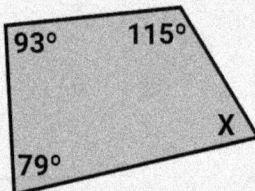

93° 115°

X

79°

x = _____

20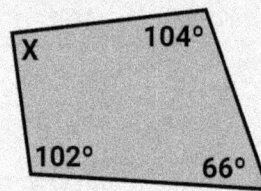

X 104°

102°

66°

x = _____

Points, Lines, and Planes

Use the image below to answer questions 1 – 6:

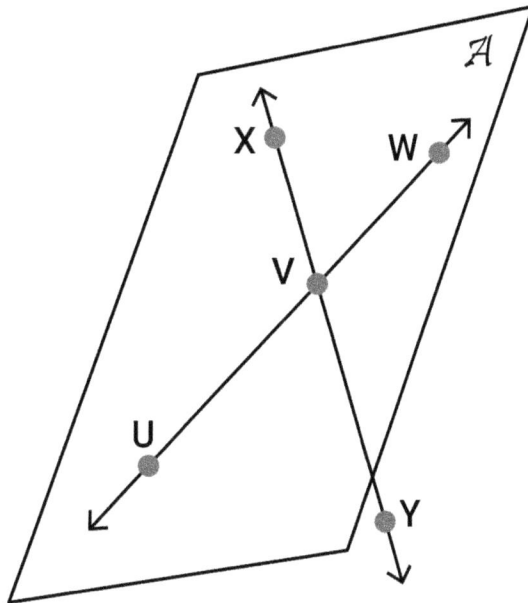

1 Name any 3 points _____ _____ _____

2 Name 2 line segments _____ _____

3 Name a set of 3 points
that are collinear and coplanar _____ _____ _____

4 Name 2 lines _____ _____

5 Name a point that is not coplanar with X and V _____

6 Name 2 rays _____ _____

Points, Lines, and Planes

Use the image below to answer questions 7 – 12:

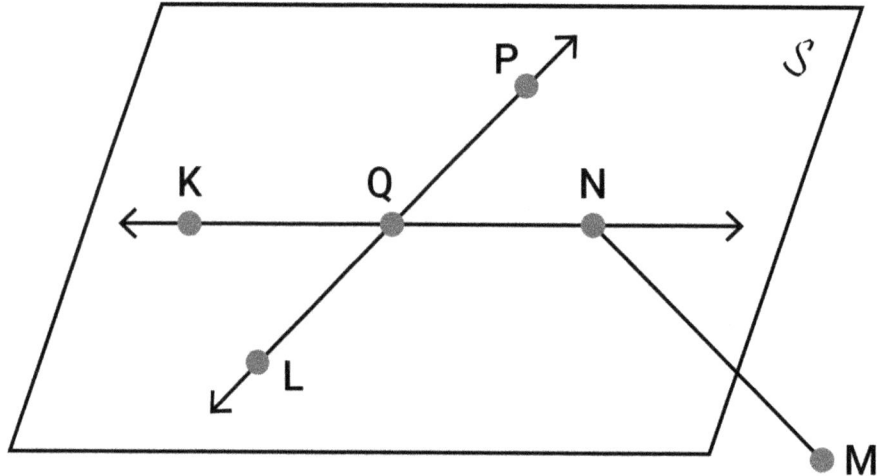

7

Name a point that is not collinear with L but is coplanar _____

8

Name a line segment that could not also be named as a ray or line _____

9

Name 2 points that are collinear with L _____ _____

10

Name 2 lines _____ _____

11

Name a point that is not coplanar with P and Q _____

12

Name 2 rays _____ _____

Points, Lines, and Planes

Use the image below to answer questions 13 – 20:

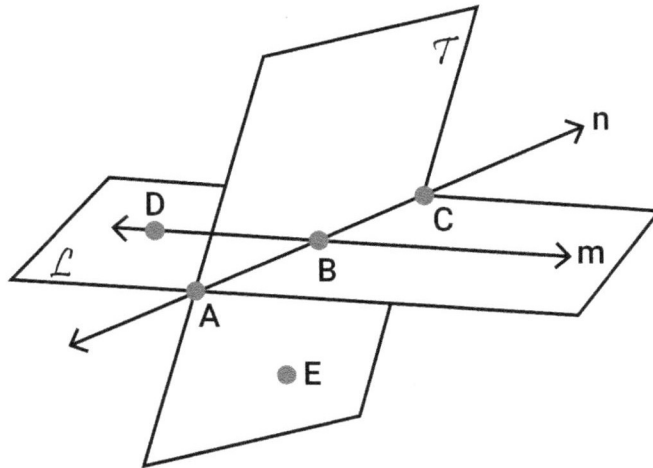

13

Name 2 planes _____ _____

14

Name a point that is collinear with D _____

15

Name a point that is not collinear with C _____

16

Write another way to name line n _____

17

Name a pair of points that
are not collinear in plane T _____ _____

18

Name the line where L and T intersect _____

19

Name a point that is coplanar with D _____

Parallel and Intersecting Lines

Identify whether the indicated lines are parallel, perpendicular, or intersecting.

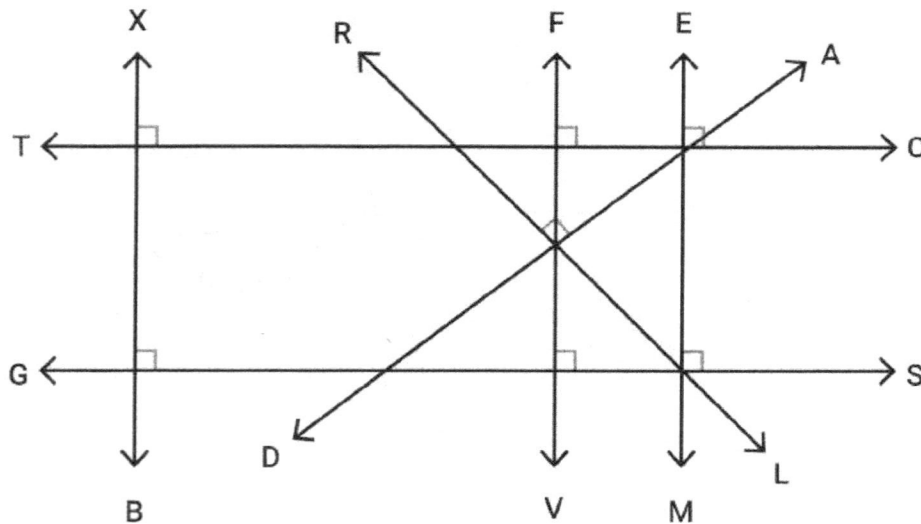

1 Lines EM and FV are _____ lines.

2 Lines GS and TC are _____ lines.

3 Lines AD and LR are _____ lines.

4 Lines BX and EM are _____ lines.

5 Lines AD and GS are _____ lines.

6 Lines BX and FV are _____ lines.

7 Lines EM and AD are _____ lines.

8 Lines BX and GS are _____ lines.

9 Lines TC and FV are _____ lines.

10 Lines RL and FV are _____ lines.

Calculating Perimeter

Determine the perimeter of the shapes.

1 A square with a side length of 8 inches.

2 A circle with a radius of 5 cm.

3 One triangle has side lengths of 4 inches, 9 inches, and 8 inches, while another equilateral triangle has a side length of 6 inches. Which triangle has a larger perimeter?

4 A hexagon with a side length of 7 cm.

5 A rectangular field is 37 meters wide and 84 meters long.

6 A rectangular bookmark that is 4 cm. wide and 12 cm. long.

7 A square with a side length of 15 mm.

8 A rhombus with a side length of 5 inches.

9 A trapezoid with side lengths of 19, 17, 15, and 17 yards.

10 A circle with a diameter of 14 feet.

Calculating Perimeter

Determine the perimeter of the shapes.

11 A regular hexagon with a side length of 12 cm.

12 A triangle with sides of 15, 17, and 18 inches.

13 A regular hexagon with a side length of 7 cm.

14 A regular octagon with a side length of 8 feet.

15 A regular pentagon with a side length of 3 yards.

16 A parallelogram with opposite sides of 13 and 19 inches.

17 A trapezoid with bases of 7 and 9 and sides of 4 inches.

18 A rectangle whose length is twice the width, which is 8 cm.

19 A triangle with two sides that are 14 inches and one side that is half as long.

20 A rectangle whose width is one-third the length, which is 27 cm.

Finding the Missing Side Lengths

Calculate the missing side lengths from the information provided.

1 The perimeter of a rectangular paper is 32 inches. One side is 7 inches.

2 The perimeter of a square piece of origami paper is 36 inches. What is the side length?

3 The perimeter of a square napkin is 20 inches. What is the side length?

4 The perimeter of a rectangular mirror is 58 inches. One side is 14 inches.

5 The perimeter of a rectangular placemat is 36 inches. One side is 10 inches.

6 The perimeter of a triangular main sail is 65 yards. One side is 29 yards and one is 19 yards.

7 The perimeter of a square playpen is 148 feet. What is the side length?

8 The perimeter of a triangle is 19 cm. One side is 7 and one is 8.

9 The perimeter of a regular hexagon is 48 centimeters. What is the length of a side?

10 A triangular chip has a perimeter of 17 centimeters. One side is 5 centimeters and another side is 6 centimeters.

Finding the Missing Side Lengths

Calculate the missing side lengths from the information provided.

11 A stop sign (an octagon) has a perimeter of 96 inches. What is the length of a side?

12 The perimeter of a rectangular pasture is 122 meters. One side is 35 meters.

13 The perimeter of a rectangular rug is 90 inches. One side is 18 inches.

14 The perimeter of a rectangular chalkboard is 140 inches. One side is 47 inches.

15 A regular pentagon has a perimeter of 45 centimeters. What is the side length?

16 The perimeter of a triangular bandana is 42 cm. One side is 17 cm and one is 13 cm.

17 The perimeter of a pentagon drawn on your paper is 85 millimeters. Four of the side lengths are as follows (in mm): 17, 11, 19, 18. What is the length of the other side?

18 The perimeter of a square cat litter box is 120 centimeters. What is the side length?

19 The perimeter of a triangle is 26 cm. One side is 9 and one is 12.

20 A six-sided polygon has a perimeter of 87 in. The sides (except 1) are as follows: 14, 13, 18, 22, 16.

Area of Triangles

Calculate the area of the following triangles.
Remember that the area of a triangle is calculated using the formula

$$A = \frac{1}{2}bh$$

1) a = 4 yards, b = 3 yards, c = 5 yards

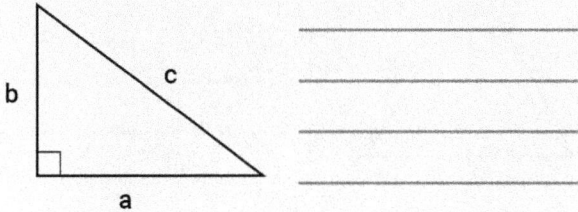

2) a = 6 cm, b = 7 cm, c = 8.5 cm, h = 3 cm

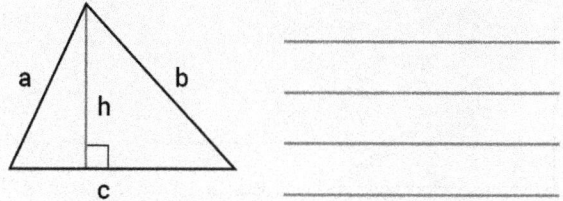

3) a = 52.72 feet, b = 90.07 feet, c = 98 feet, h = 48 feet

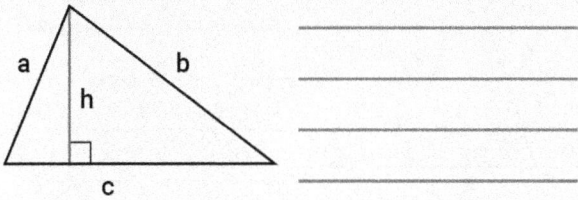

4) a = 5 feet, b = 12 feet, c = 13 feet

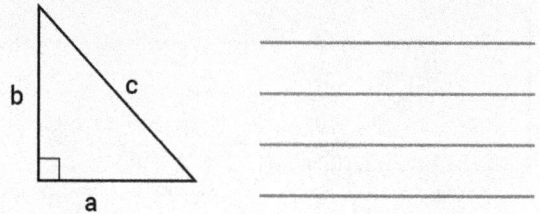

5) a = 41 meters, b = 65 meters

6) s = 4 feet

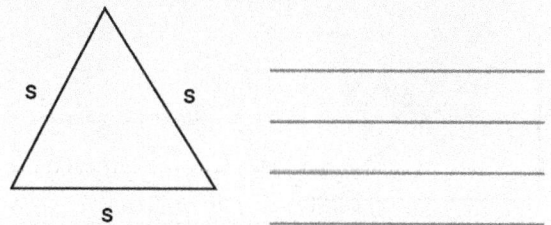

7) a = 24 cm, b = 7 cm, c = 25 cm

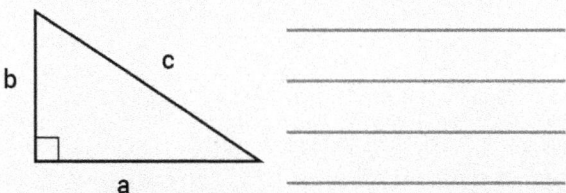

8) a = 4 yards, b = 5 yards

Area of Triangles

Calculate the area of the following triangles.
Remember that the area of a triangle is calculated using the formula

$$A = \frac{1}{2}bh$$

9   a = 16 mm, b = 12 mm, c = 20 mm

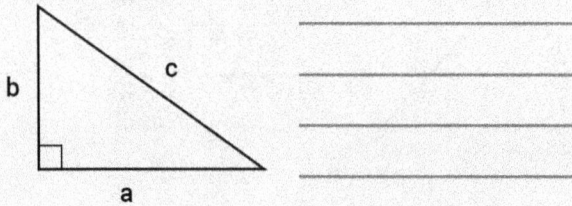

10   s = 8 inches

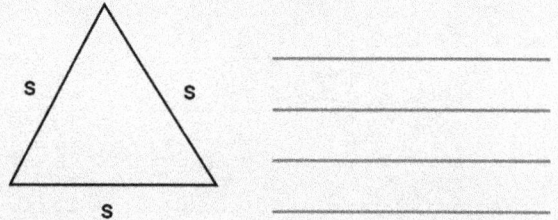

11   a = 84 mm, b = 56 mm, c = 100.96 mm

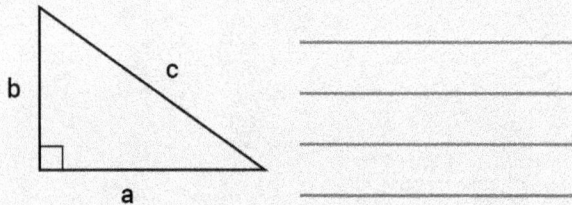

12   a = 10 feet, b = 12 feet

13   a = 70 yards, b = 56 yards, c = 89.64 yards

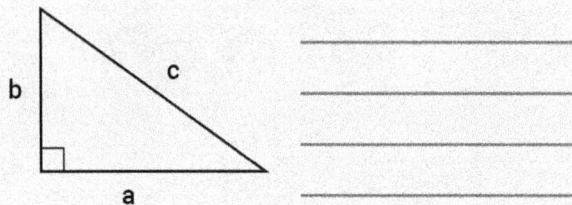

14   a = 75 cm, b = 55 mm, c = 93.01 mm

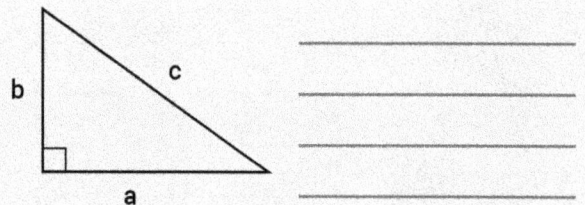

15   a = 72 inches, b = 58 inches

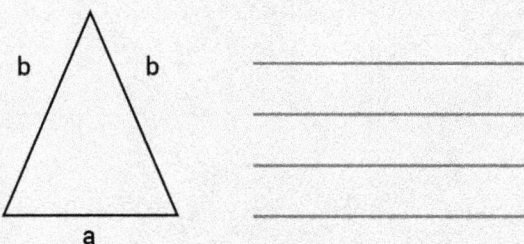

Area of Circles

Calculate the area of the following circles.
Recall that the circumference of a circle is found by calculating πr^2.
For the first column, report the exact measurement (using π).
For the second column, use 3.14 for π. An example is shown below.

Example: *A circle with a radius of 3 yards.*

	Exact Circumference	Approximate Circumference
	9π square yards	28.26 sq. yards

1 A circle with a radius of 2 meters.

Exact Circumference	Approximate Circumference
_____	_____

2 A circle with a radius of 1 inch.

Exact Circumference	Approximate Circumference
_____	_____

3 A circle with a diameter of 12 feet.

Exact Circumference	Approximate Circumference
_____	_____

4 A circle with a diameter of 26mm.

Exact Circumference	Approximate Circumference
_____	_____

5 A circle with a diameter of 18 feet.

Exact Circumference	Approximate Circumference
_____	_____

6 A circle with a radius of 10 meters.

Exact Circumference	Approximate Circumference
_____	_____

7 A circle with a radius of 5 centimeters.

Exact Circumference	Approximate Circumference
_____	_____

8 A circle with a radius of 8 inches.

Exact Circumference	Approximate Circumference
_____	_____

9 A circle with a diameter of 14 inches.

Exact Circumference	Approximate Circumference
_____	_____

10 A circle with a diameter of 24 feet.

Exact Circumference	Approximate Circumference
_____	_____

Area of Mixed Shapes and Figures

Determine the area of the following shapes and figures

1 A square with a side length of 3 meters.

_____
_____
_____
_____

2 A rectangle that is 11 feet by 5 feet.

_____
_____
_____
_____

3 A rectangle that is 8 inches by 7 inches.

_____
_____
_____
_____

4 A square with a side length of 6 centimeters.

_____
_____
_____
_____

5 A triangle with a base of 4 inches and a height of 7 inches.

_____
_____
_____
_____

6 A triangle with a base of 12 cm and a height of 8 cm.

_____
_____
_____
_____

7 A square with a side length of 15 inches.

_____
_____
_____
_____

8 A rectangle with a length of 12 inches and a width of 5 inches.

_____
_____
_____
_____

9 A circle with a radius of 9 cm.

_____
_____
_____
_____

10 A triangle with a base of 4 inches and a height of 5 inches.

_____
_____
_____
_____

Area of Mixed Shapes and Figures

Determine the area of the following shapes and figures

11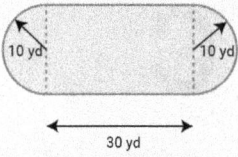

10 yd   10 yd

30 yd

_____
_____
_____
_____

12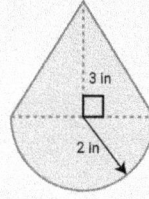

3 in

2 in

_____
_____
_____
_____

13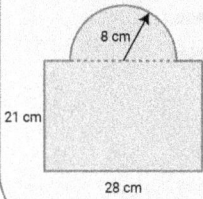

8 cm

21 cm

28 cm

_____
_____
_____
_____
_____

14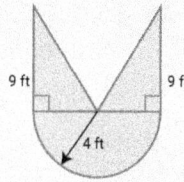

9 ft   9 ft

4 ft

_____
_____
_____
_____
_____

15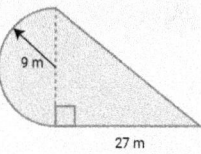

9 m

27 m

_____
_____
_____
_____
_____

16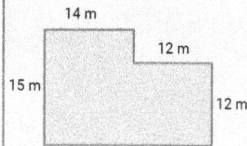

14 m

12 m

15 m

12 m

_____
_____
_____
_____
_____

17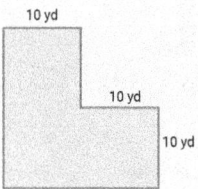

10 yd

10 yd

10 yd

_____
_____
_____
_____
_____

18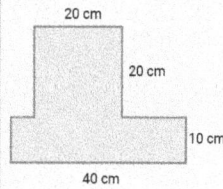

20 cm

20 cm

10 cm

40 cm

_____
_____
_____
_____
_____

19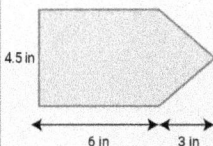

4.5 in

6 in   3 in

_____
_____
_____
_____
_____

20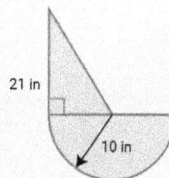

21 in

10 in

_____
_____
_____
_____
_____

Lines of Symmetry in the Alphabet

For each capital letter, determine if there is one or more lines of symmetry. If so, draw them in. If not, write "no symmetry."

1. **A**
   _____

2. **B**
   _____

3. **C**
   _____

4. **D**
   _____

5. **E**
   _____

6. **F**
   _____

7. **G**
   _____

8. **H**
   _____

9. **I**
   _____

10. **J**
    _____

11. **K**
    _____

12. **L**
    _____

13. **M**
    _____

14. **N**
    _____

15. **O**
    _____

Lines of Symmetry in the Alphabet

For each capital letter, determine if there is one or more lines of symmetry. If so, draw them in. If not, write "no symmetry."

16) P

_____

17) Q

_____

18) R

_____

19) S

_____

20) T

_____

21) U

_____

22) V

_____

23) W

_____

24) X

_____

25) Y

_____

26) Z

_____

Reading the Coordinate Plane

Use the provided coordinate planes with points plotted to write the coordinates of the labeled points.

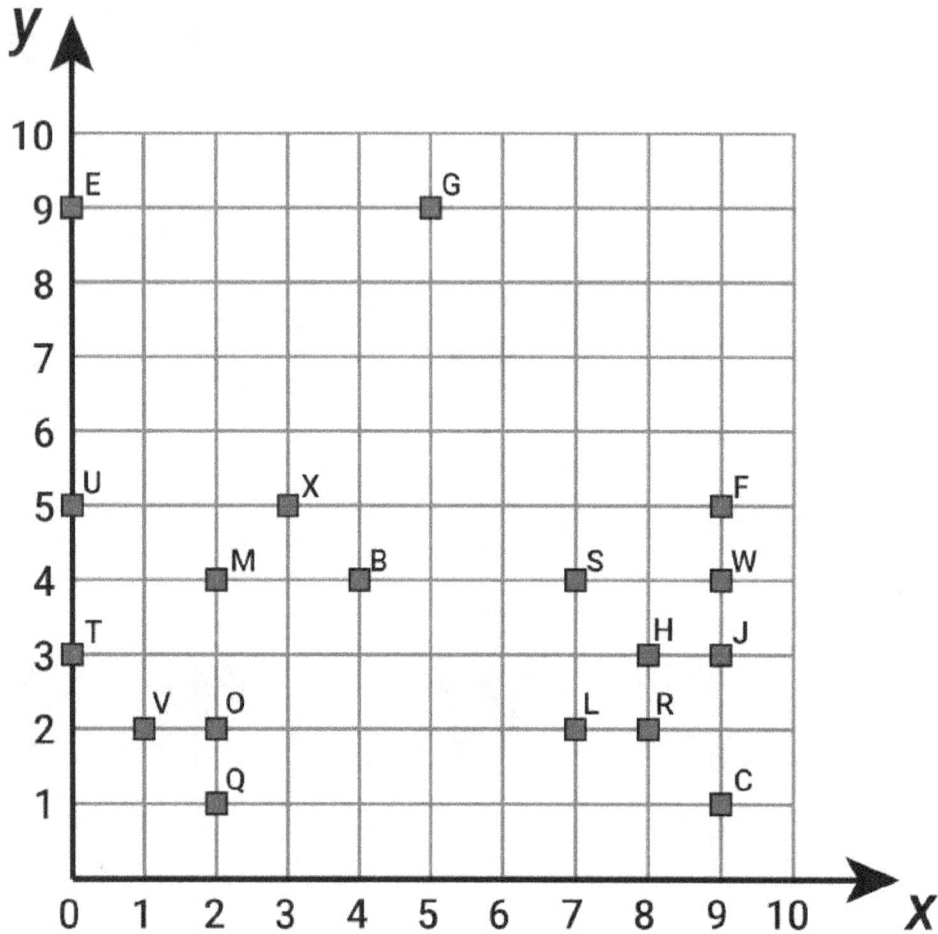

1. B _____
2. C _____
3. E _____
4. F _____
5. G _____
6. H _____
7. J _____
8. L _____
9. M _____
10. O _____
11. Q _____
12. R _____
13. S _____
14. T _____
15. U _____
16. V _____
17. W _____
18. X _____

Reading the Coordinate Plane

Use the provided coordinate planes with points plotted to write the coordinates of the labeled points.

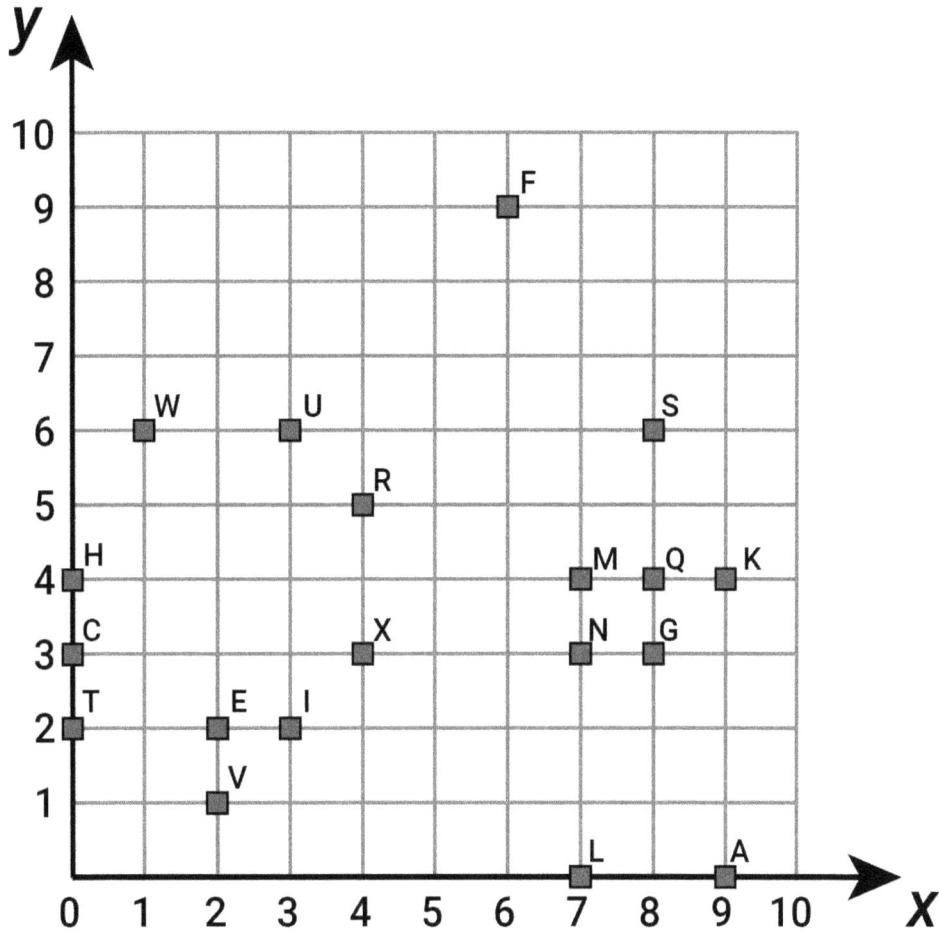

1 A _____

2 C _____

3 E _____

4 F _____

5 G _____

6 H _____

7 I _____

8 K _____

9 L _____

10 M _____

11 N _____

12 Q _____

13 R _____

14 S _____

15 T _____

16 U _____

17 V _____

18 W _____

Plotting Points in the Coordinate Plane

Plot each set of points on the blank coordinate plane.

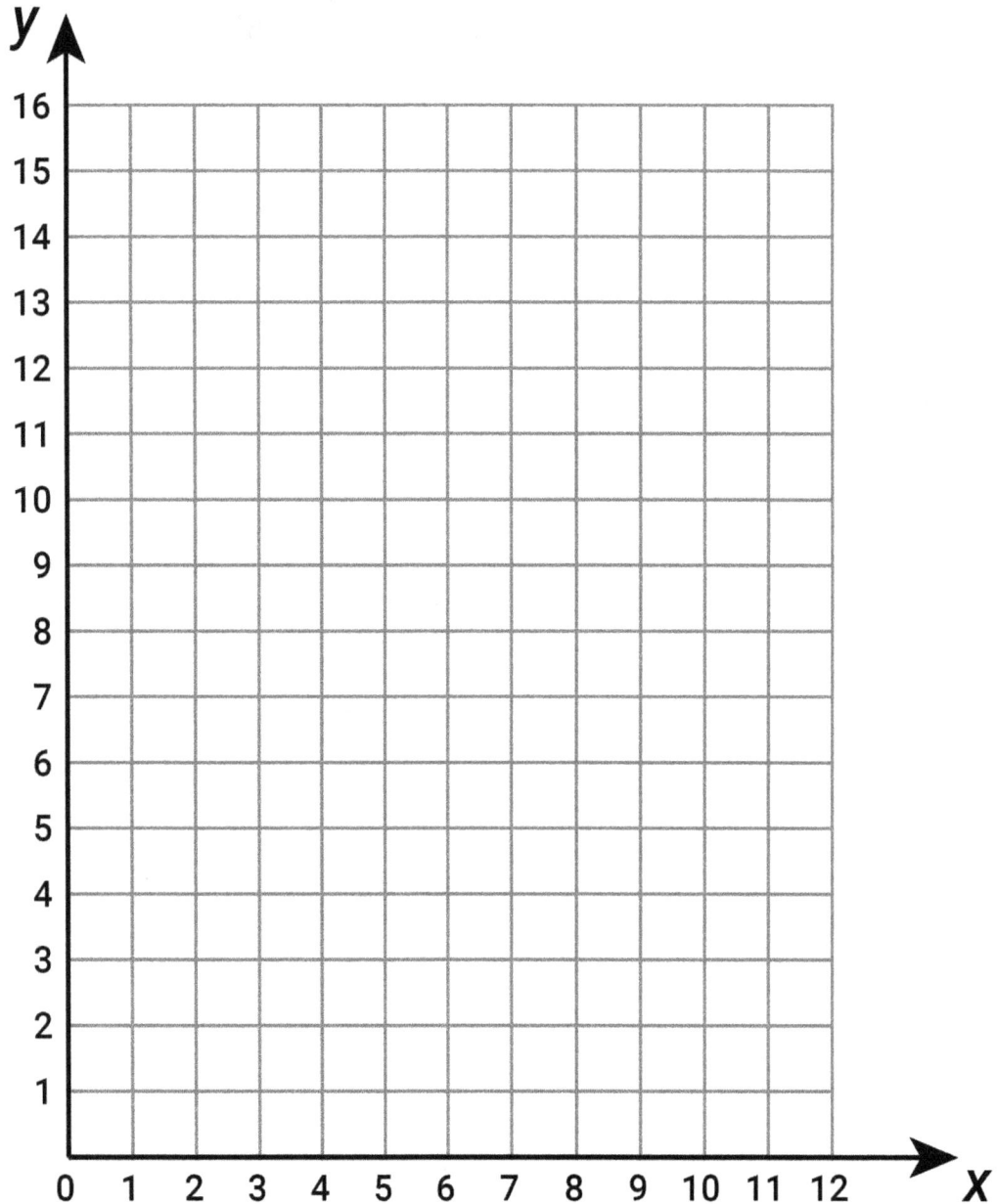

1 A (3, 1)   **2** B (2, 3)   **3** C (4, 8)   **4** D (6, 9)   **5** F (7, 7)

6 G (0, 3)   **7** H (6, 2)   **8** I (8, 2)   **9** K (6, 3)   **10** L (6, 0)

11 M (8, 5)   **12** P (1, 5)   **13** R (1, 8)   **14** S (0,2)   **15** T (0, 9)

16 V (5, 8)   **17** X (0, 7)   **18** Y (1, 4)   **19** Z (8, 6)

Plotting Points in the Coordinate Plane

Plot each set of points on the blank coordinate plane.

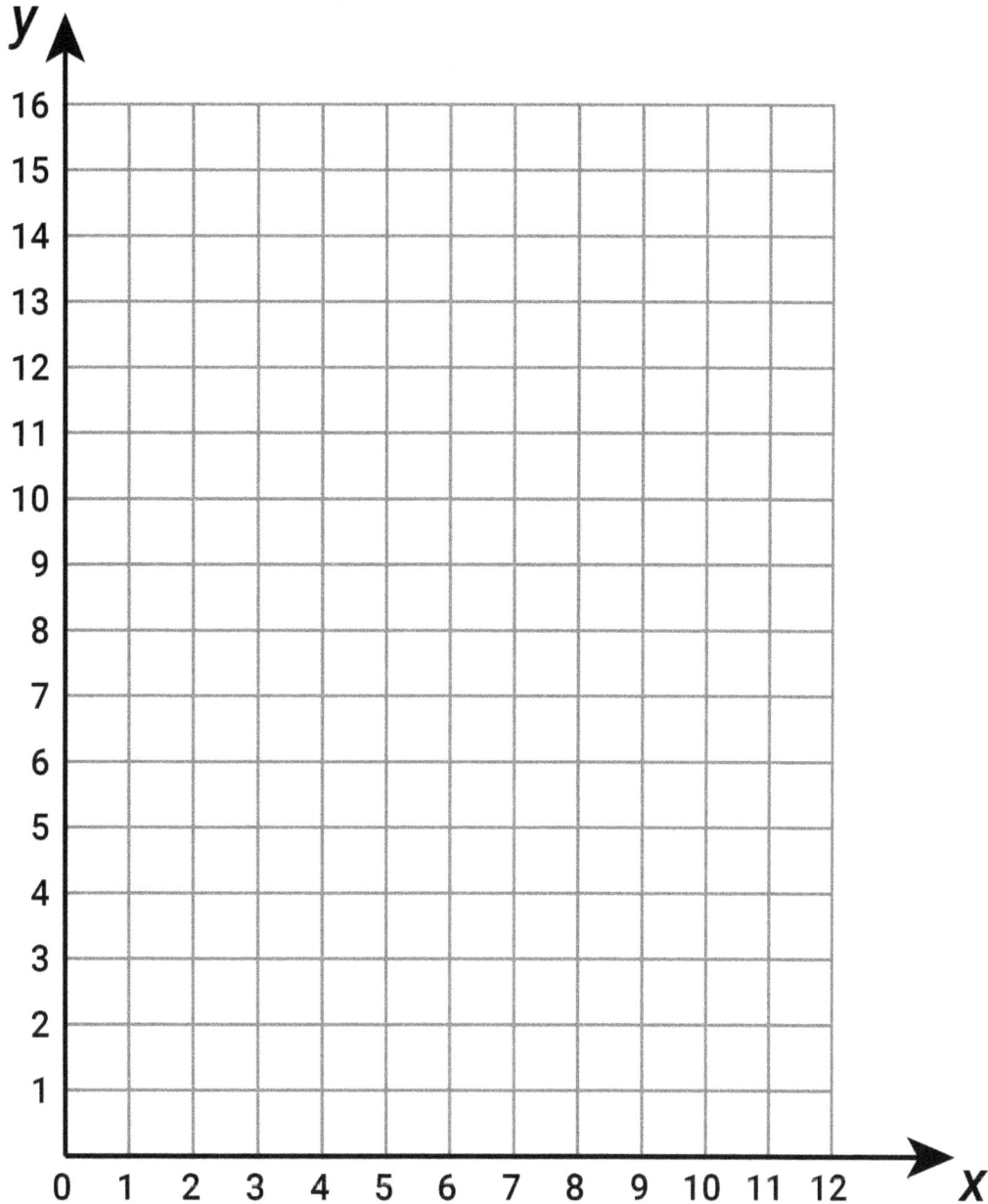

1. B (4, 6)
2. C (5, 0)
3. D (2, 9)
4. E (8, 9)
5. F (3, 8)
6. G (7, 4)
7. H (5, 5)
8. J (0, 3)
9. K (3, 2)
10. L (5, 3)
11. M (8, 4)
12. N (1, 9)
13. O (2, 1)
14. P (5, 9)
15. R (5, 8)
16. S (7, 5)
17. T (6, 8)
18. V (0, 0)
19. X (9, 8)

Identifying Shapes

Write the name of the shape next to each shape below.

13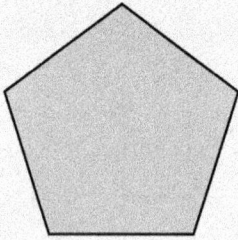

Name: _____

14

Name: _____

15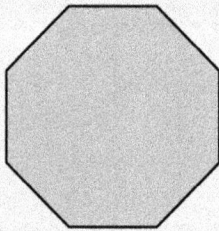

Name: _____

16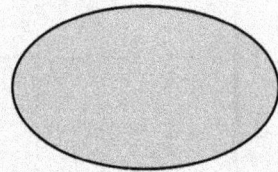

Name: _____

17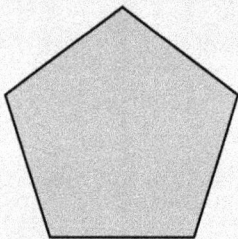

Name: _____

18

Name: _____

19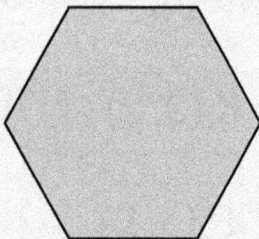

Name: _____

20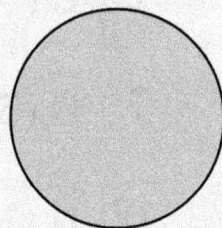

Name: _____

Identifying Shapes

Write the name of the shape next to each shape below.

7

Name: _____

8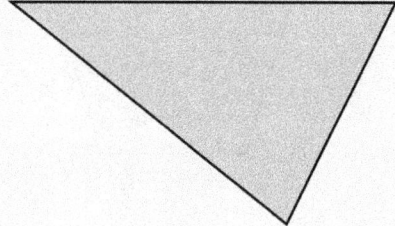

Name: _____

9

Name: _____

10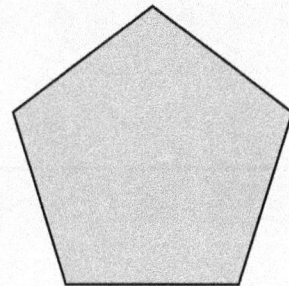

Name: _____

11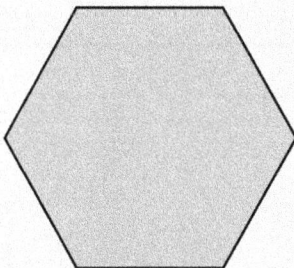

Name: _____

12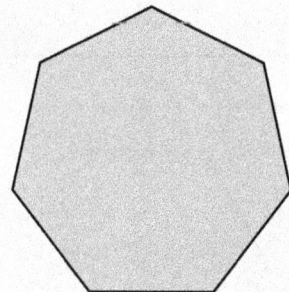

Name: _____

Identifying Shapes

Write the name of the shape next to each shape below.

13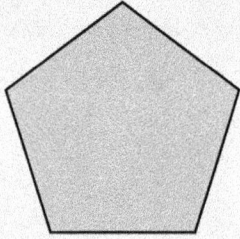

Name: _____

14

Name: _____

15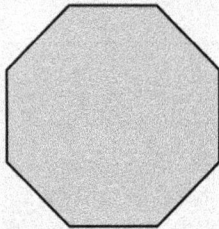

Name: _____

16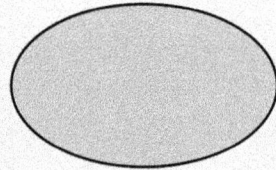

Name: _____

17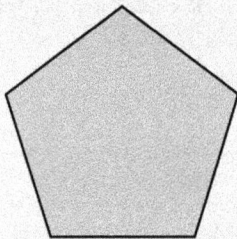

Name: _____

18

Name: _____

19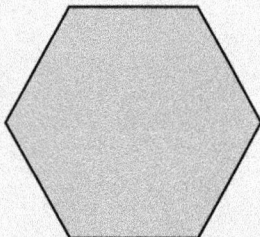

Name: _____

20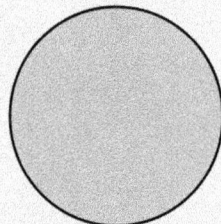

Name: _____

Drawing Shapes with Specific Attributes

Fill in the chart below by drawing the specified shape and then filling
in the number of sides and vertices. To challenge yourself,
try to make notes in the notes box about specifics of the shape.
For example, if it's a square, you might write "four sides of equal length"
because this helps differentiate a square from another four-sided figure.
The shapes will be in a random order.

Shape	Drawing	Number of Sides and Vertices	Notes
Circle			
Pentagon			
Square			
Octagon			
Triangle			
Rectangle			

Drawing Shapes with Specific Attributes

Fill in the chart below by drawing the specified shape and then filling
in the number of sides and vertices. To challenge yourself,
try to make notes in the notes box about specifics of the shape.
For example, if it's a square, you might write "four sides of equal length"
because this helps differentiate a square from another four-sided figure.
The shapes will be in a random order.

Shape	Drawing	Number of Sides and Vertices	Notes
Trapezoid			
Rhombus			
Hexagon			
Oval			
Parallelogram			
Septagon/ Heptagon			

Classifying Triangles

Look at the images of the following triangles and classify them as equilateral, scalene, or isosceles based on their side lengths.

1
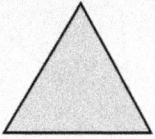
_____

2
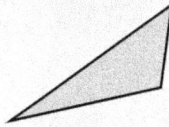
_____

3
_____

4
_____

5
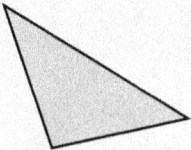
_____

6
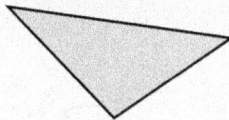
_____

7
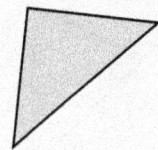
_____

8
_____

9
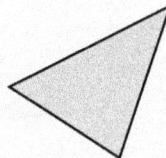
_____

10
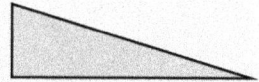
_____

11
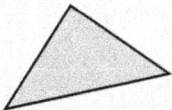
_____

12
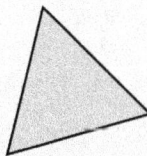
_____

13
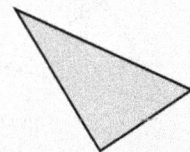
_____

14

_____

15
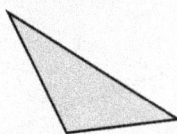
_____

Classifying Triangles

For this next set of images, classify the triangles
as acute, obtuse, or right based on their angles.

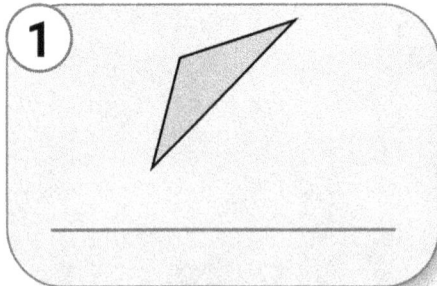

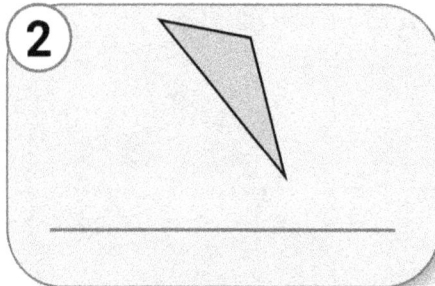

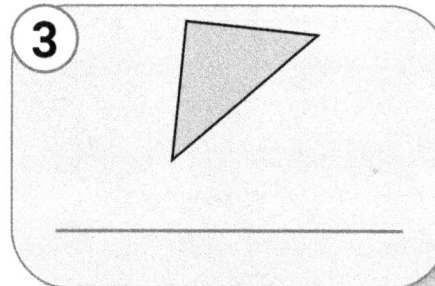

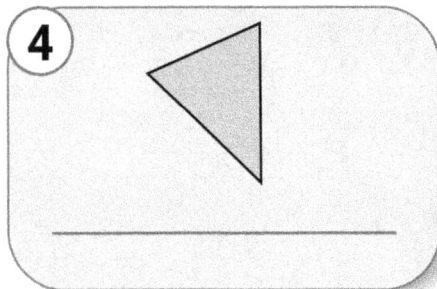

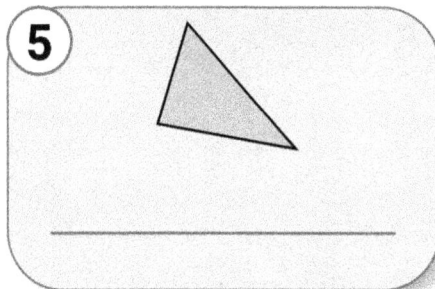

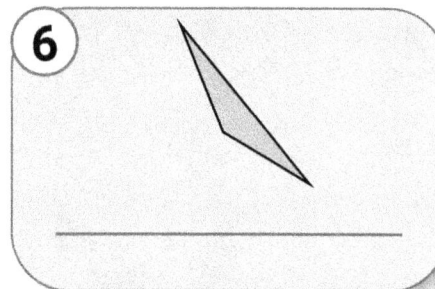

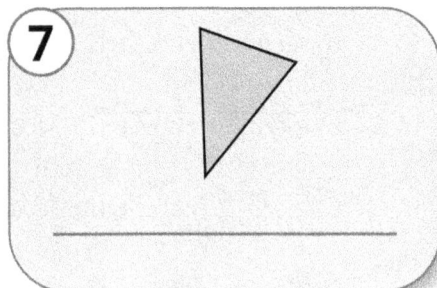

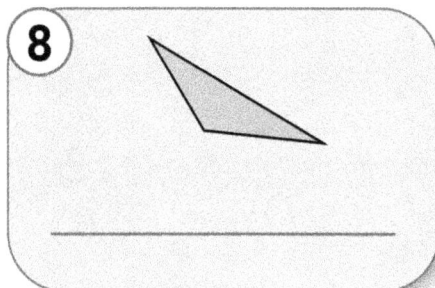

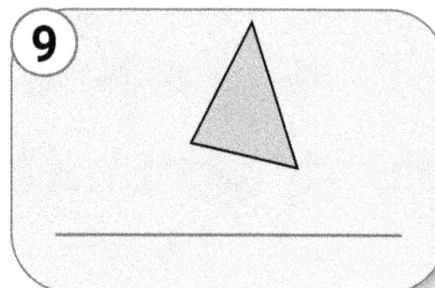

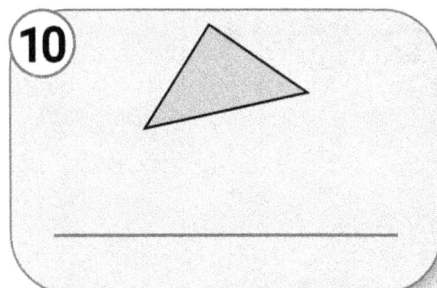

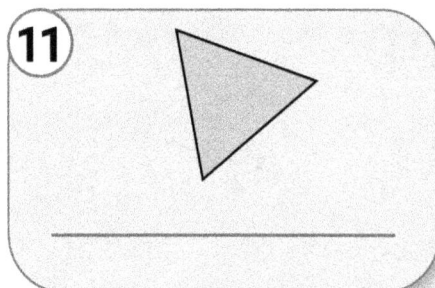

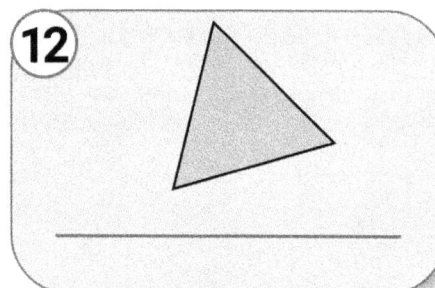

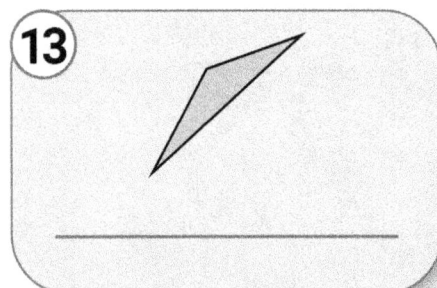

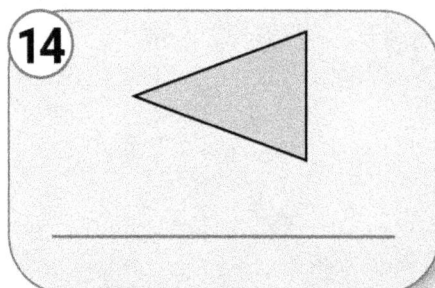

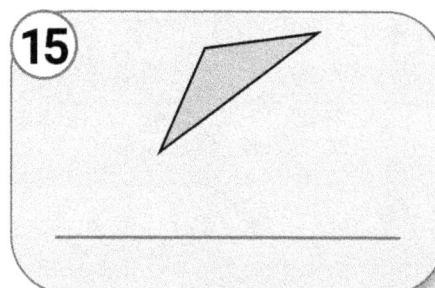

1

_____

2

_____

3

_____

4

_____

5

_____

6

_____

7

_____

8

_____

9

_____

10

_____

11

_____

12

_____

13

_____

14

_____

15

_____

Classifying Triangles

Now, put it together and classify the following triangles by their sides and angles (for example, isosceles and right).

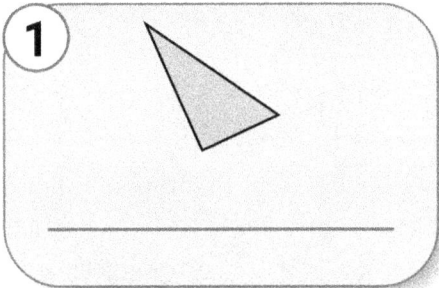

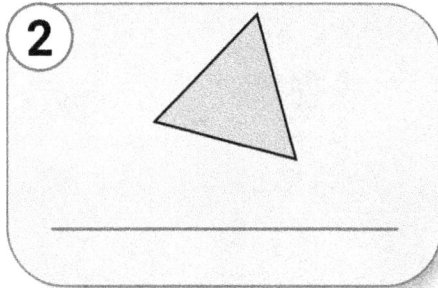

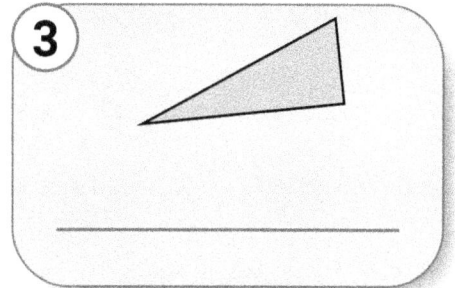

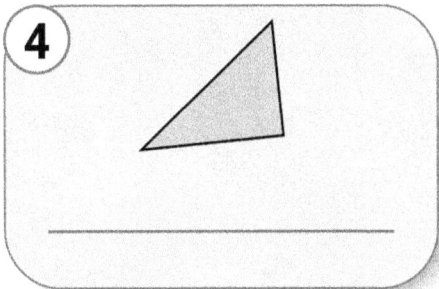

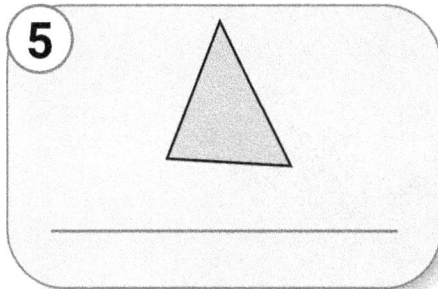

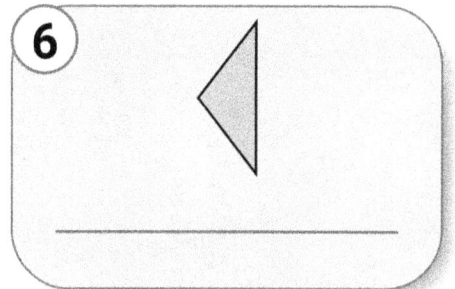

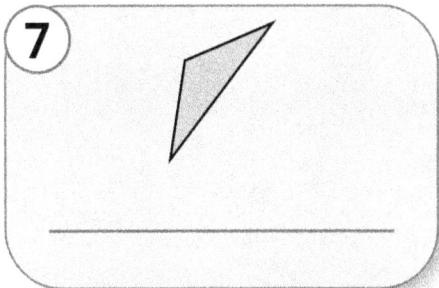

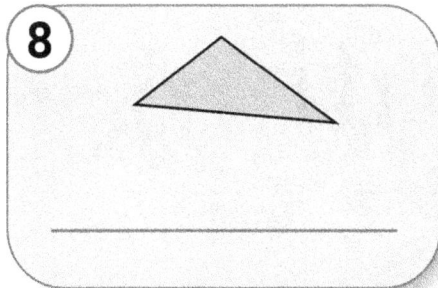

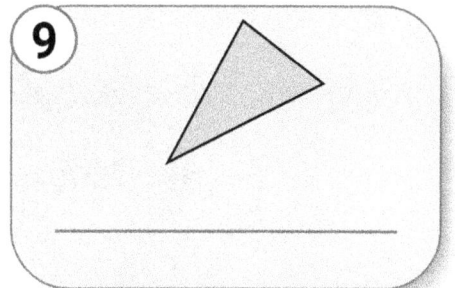

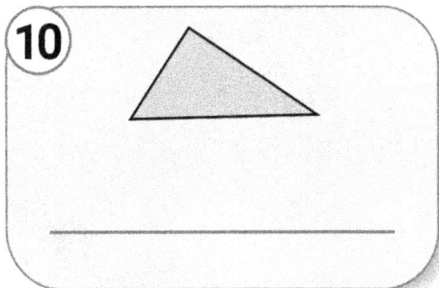

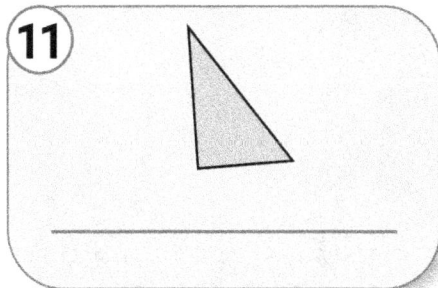

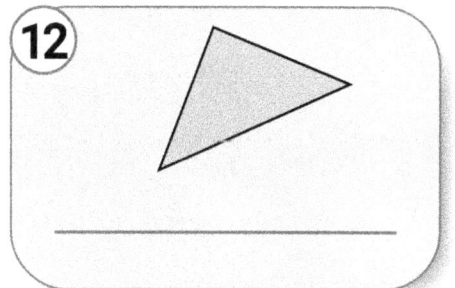

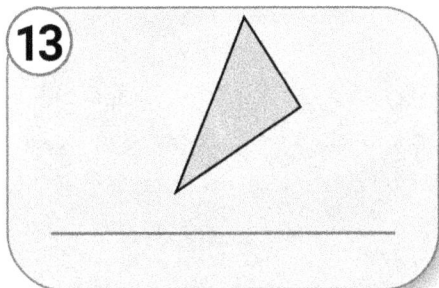

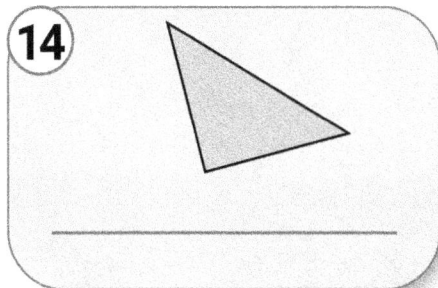

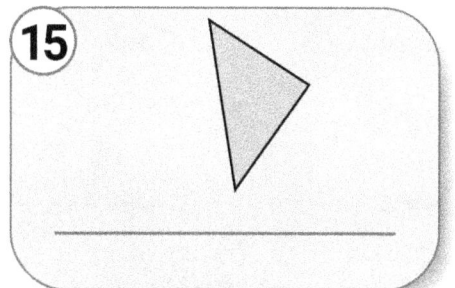

1 _____

2 _____

3 _____

4 _____

5 _____

6 _____

7 _____

8 _____

9 _____

10 _____

11 _____

12 _____

13 _____

14 _____

15 _____

Arithmetic

Identifying Place Value

1. Tens
2. Thousands
3. Ten thousands
4. Hundreds
5. Ones
6. Hundred thousands
7. Thousands
8. Tens
9. Ten thousands
10. Ones
11. Ten thousands
12. Hundred thousands
13. Hundreds
14. Millions
15. Ten thousands

Identifying the Number of a Given Place Value

1. Hundreds
2. Tens
3. Hundred thousands
4. Ones
5. Hundreds
6. Ten thousands
7. Thousands
8. Tens
9. Millions
10. Ten thousands
11. Tenths
12. Hundred thousands
13. Ten thousands
14. Ones
15. Hundredths

Compare the Values

1. =
2. =
3. =
4. <
5. =
6. <
7. <
8. >
9. =
10. =

Decomposing Numbers

1 798   7 hundreds, 9 tens, 8 ones

2 501   5 hundreds, 0 tens, 1 ones

3 927.05   9 hundreds, 2 tens, 7 ones, 0 tenths, 5 hundredths

4 260.23   2 hundreds, 6 tens, 0 ones, 2 tenths, 3 hundredths

5 2,709   2 thousands, 7 hundreds, 0 tens, 9 ones

6 6,813   6 thousands, 8 hundreds, 1 tens, 3 ones

7 1,982   1 thousands, 9 hundreds, 8 tens, 2 ones

8 8,714   8 thousands, 7 hundreds, 1 tens, 4 ones

9 11,920   1 ten thousands, 1 thousands, 9 hundreds, 2 tens, 0 ones

10 90,101   9 ten thousands, 0 thousands, 1 hundreds, 0 tens, 1 ones

11 45,341   4 ten thousands, 5 thousands, 3 hundreds, 4 tens, 1 ones

12 29,800   2 ten thousands, 9 thousands, 8 hundreds, 0 tens, 0 ones

13 100,954   1 hundred thousands, 0 ten thousands, 0 thousands, 9 hundreds, 5 tens, 4 ones

14 302,540   3 hundred thousands, 0 ten thousands, 2 thousands, 5 hundreds, 4 tens, 0 ones

15 854,624   8 hundred thousands, 5 ten thousands, 4 thousands, 6 hundreds, 2 tens, 4 ones

Writing Numbers from Their Components

1. 672
2. 760
3. 153
4. 1,002
5. 1,439
6. 582
7. 319
8. 704
9. 126
10. 876
11. 940.13
12. 497.004
13. 8,201
14. 658.9
15. 300
16. 14,941
17. 80,667
18. 0.184
19. 810
20. 475

Reading and Writing Numbers

1. Nine thousand seven hundred forty-seven
2. Six hundred thirty-nine thousand, thirty-two
3. Eighty-one thousand, three hundred
4. Two million, three hundred twenty-four thousand, three hundred seventy-one
5. Eight hundred one million, six hundred ninety thousand, eight hundred two
6. Five thousand three hundred ninety-four and sixty-three hundredths
7. One hundred twenty thousand six hundred thirty-eight and five tenths
8. Seven thousand four hundred seventy and three thousandths
9. Ninety-seven thousand nine and seven thousand one hundred eleven ten-thousandths
10. Eight hundredths
11. 1,000
12. 57
13. 112
14. 800
15. 475
16. 702
17. 580
18. 976
19. 290
20. 329
21. 701
22. 616
23. 888
24. 409
25. 932

Practice Makes Perfect: Addition and Subtraction Review

1) 160

2) 114

3) 103

4) 1542

5) 1371

6) 1517

7) 1751

8) 11033

9) 14104

10) 10830

11) 89

12) 32

13) 833

14) 188

15) 6852

16) 8457

17) 4967

18) 331

19) 45479

20) 15892

21) 772

22) 588

23) 182

24) 222

25) 863

26) 507

27) 588

28) 436

29) 188

30) 335

31) 304

32) 858

33) 440

34) 474

35) 232

36) 641

37) 427

38) 705

39) 205

40) 231

41) 233

42) 322

43) 787

44) 275

45) 307

46) 799

47) 697

48) 350

49) 561

50) 383

51) 983

52) 780

53) 509

54) 178

55) 356

56) 471

57) 116

58) 286

59) 26

60) 267

61) 446

62) 636

63) 736

64) 416

65) 163

66) 812

67) 239

68) 403

69) 792

70) 895

Real-World Addition and Subtraction Problems

1 15 + 7 = 22 animals

2 31 − 8 = 23. 23 − 7 = 16 pounds

3 9 + 9 = 18 dumplings

4 6 + 5 = 11 rooms

5 8 + 6 = 14 drawing utensils

6 2 + 1 + 4 = 7 pets

7 2 + 2 + 3 + 4 + 3 = 14 family members

8 2 + 1 + 5 + 4 + 1 = 13 balls

9 5 + 4 = 9 weeks

10 20 − 5 − 7 = 8 bracelets

11 23 − 12 = 11 girls

12 4 + 3 + 5 + 2 = 14 films

13 3 + 6 + 3 = 12 items

14 5 + 5 + 5 + 2 = 17 dishes

15 8 + 3 + 5 + 2 = 18 pizzas

16 4 + 8 + 3 + 2 = 17 bagels

17 15 gifts

18 18 minutes

19 16 cookies

20 20 − 13 = 7 tickets

21 7 + 4 = 11 action figures

22 18 minutes

23 16 − 5 = 11 cars

24 27 raisins

25 23 students

26 35 students

27 30 students

28 30 items

29 30 − 15 = 15 minutes

30 $7

Multiplying a 1-Digit Number by a 1-Digit Number

(1) 18	(2) 35	(3) 24	(4) 24	(5) 32
(6) 6	(7) 63	(8) 0	(9) 8	(10) 15
(11) 40	(12) 24	(13) 27	(14) 12	(15) 2
(16) 0	(17) 21	(18) 20	(19) 5	(20) 18
(21) 5	(22) 81	(23) 0	(24) 49	(25) 48

Multiplying a 2-Digit Number by a 1-Digit Number

(1) 80	(2) 192	(3) 450	(4) 36	(5) 392
(6) 576	(7) 344	(8) 413	(9) 0	(10) 92
(11) 657	(12) 402	(13) 396	(14) 216	(15) 0
(16) 666	(17) 0	(18) 408	(19) 558	(20) 304
(21) 408	(22) 640	(23) 22	(24) 29	(25) 0

Multiplying a 2-Digit Number by a 2-Digit Number

1) 4940
2) 860
3) 4758
4) 1584
5) 4758
6) 1404
7) 4785
8) 893
9) 1188
10) 1458
11) 990
12) 7821
13) 1005
14) 748
15) 5082
16) 2550
17) 3081
18) 720
19) 456
20) 3196
21) 2205
22) 2100
23) 1176
24) 2584
25) 2736

Multiplying a 3-Digit Number by a 2-Digit Number

1) 44640
2) 77163
3) 6156
4) 30804
5) 56034
6) 6048
7) 16575
8) 40736
9) 29328
10) 15756
11) 11960
12) 13826
13) 10296
14) 7992
15) 4862
16) 24765
17) 30690
18) 34960
19) 33176
20) 35259
21) 22723
22) 36894
23) 11172
24) 38808
25) 21340

Dividing By 1 – 10

1. 3
2. 10
3. 8
4. 7
5. 6
6. 4
7. 5
8. 7
9. 3
10. 3
11. 7
12. 1
13. 9
14. 5
15. 4
16. 42
17. 10
18. 2
19. 10
20. 2
21. 6
22. 3
23. 10
24. 4
25. 5

Dividing By 1 – 100

1. 98
2. 100
3. 66
4. 95
5. 5
6. 11
7. 27
8. 18
9. 34
10. 50
11. 79
12. 42
13. 69
14. 93
15. 56
16. 36
17. 83
18. 87
19. 7
20. 20
21. 55
22. 1
23. 52
24. 62
25. 74

Dividing a 3-Digit Number by a 2-Digit Number

1 11	**2** 95	**3** 7	**4** 7	**5** 13
6 8	**7** 81	**8** 72	**9** 9	**10** 9
11 81	**12** 13	**13** 11	**14** 4	**15** 18
16 20	**17** 82	**18** 37	**19** 13	**20** 14
21 6	**22** 17	**23** 8	**24** 21	**25** 8

Prime or Composite?

1 Prime

2 Composite

3 Composite

4 Prime

5 Composite

6 Composite

7 Composite

8 Prime

9 Composite

10 Composite

11 Prime

12 Prime

13 Composite

14 Prime

15 Prime

Greatest Common Factor

1 5	**2** 2	**3** 2	**4** 1	**5** 3
6 12	**7** 1	**8** 10	**9** 4	**10** 3
11 4	**12** 10	**13** 11	**14** 13	**15** 16

Least Common Multiple

1 120	**2** 30	**3** 12	**4** 15	**5** 60
6 60	**7** 30	**8** 6	**9** 40	**10** 120
11 24	**12** 40	**13** 15	**14** 30	**15** 60

Prime Factorization

1

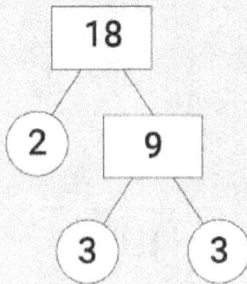

18 = 2 x 3 x 3

2

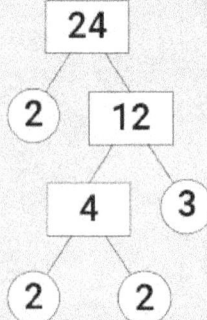

24 = 2 x 2 x 2 x3

3

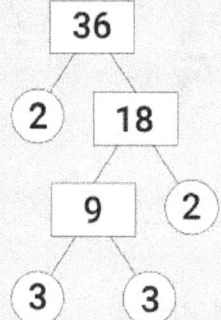

36 = 2 x 2 x 3 x 3

4

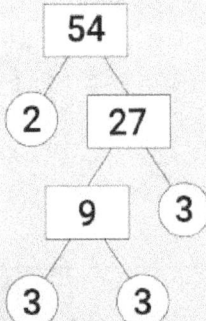

54 = 2 x 3 x 3 x 3

5

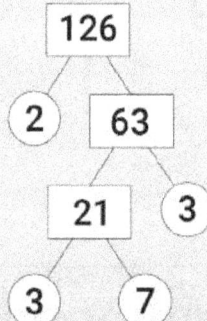

126 = 2 x 3 x 3 x 7

6

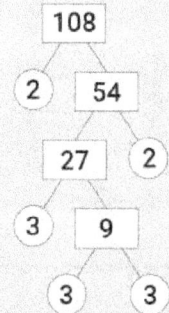

108 = 2 x 2 x 3 x 3 x 3

7

51 = 3 x 17

8

44 = 2 x 2 x 11

9

85 = 5 x 17

10

46 = 2 x 23

11

74 = 2 x 37

12

90 = 2 x 3 x 3 x 5

13

48 = 2 x 2 x 2 x 2 x 3

14

94 = 2 x 47

15

128 = 2 x 2 x 2 x 2 x 2 x 2 x 2

Fill in the Table

Complete the table by rounding to the nearest value indicated by each column.
The first one has been completed as an example.

	Hundredths	Tenths	Ones	Tens	Hundreds	Thousands
8316.728	8316.73	8316.7	8317	8320	8300	8000
528.652	528.65	528.7	529	530	500	1000
4381.748	4381.75	4381.7	4382	4380	4400	4000
1273.025	1273.03	1273.0	1273	1270	1300	1000
9924.703	9924.70	9924.7	9925	9920	9900	10000
3092.816	3092.82	3092.8	3093	3090	3100	3000
501.105	501.11	501.1	501	500	500	1000
9281.002	9281.00	9281.0	9281.0	9280	9300	9000
278.056	278.06	278.1	278	280	300	0
2244.525	2244.53	2244.5	2245	2240	2200	2000
735.642	735.64	735.6	736	740	700	1000
1387.955	1387.96	1388.0	1388	1390	1400	1000
984.058	984.06	984.1	984	980	1000	1000
3629.765	3629.77	3629.8	3630	3630	3600	4000
7445.869	7445.87	7445.9	7446	7450	7400	7000
8401.836	8401.84	8401.8	8402	8400	8400	8000
5832.547	5832.55	5832.5	5833	5830	5800	6000
675.498	675.50	675.50	675	680	700	1000
3268.027	3268.03	3268.0	3268	3270	3300	3000
499.367	499.37	499.4	499	500	500	0

Using Rounding to Estimate the Sum or Difference

1 50 + 40 = 90

2 80 − 40 = 40

3 20 + 30 = 50

4 90 − 20 = 70

5 70 + 80 = 150

6 40 − 10 = 30

7 110 + 60 = 170

8 240 − 80 = 60

9 340 + 590 = 930

10 330 + 320 = 650

11 900 − 530 = 370

12 890 + 350 = 1240

13 700 + 550 = 1250

14 860 + 880 = 1740

15 910 + 880 = 1790

Using Rounding to Estimate the Product or Quotient

1 10 x 40 = 400

2 20 x 30 = 600

3 40 x 80 = 3200

4 90 x 80 = 7200

5 70 x 40 = 2800

6 30 x 10 = 300

7 80 x 70 = 5600

8 50 x 40 = 2000

9 20 x 30 = 600

10 90 x 40 = 3600

11 90 x 40 = 3600

12 60 x 80 = 4800

13 20 x 60 = 1200

14 80 / 20 = 4

15 100 / 20 = 5

Reducing Fractions

1 $\dfrac{3}{5}$

2 $\dfrac{3}{4}$

3 $\dfrac{1}{3}$

4 $\dfrac{1}{2}$

5 $\dfrac{7}{10}$

6 $\dfrac{3}{7}$

7 $\dfrac{1}{5}$

8 $\dfrac{1}{2}$

9 $\dfrac{1}{2}$

10 $\dfrac{1}{3}$

11 $\dfrac{1}{2}$

12 $\dfrac{3}{7}$

13 $\dfrac{2}{3}$

14 $\dfrac{1}{5}$

15 $\dfrac{1}{2}$

16 $\dfrac{8}{9}$

17 $\dfrac{2}{7}$

18 $\dfrac{1}{4}$

19 $\dfrac{2}{3}$

20 $\dfrac{3}{7}$

21 $\dfrac{9}{10}$

22 $\dfrac{1}{2}$

23 $\dfrac{2}{9}$

24 $\dfrac{3}{8}$

25 $\dfrac{5}{6}$

Reducing Fractions

1 $\dfrac{1}{8}$

2 $\dfrac{3}{8}$

3 $\dfrac{2}{11}$

4 $\dfrac{5}{7}$

5 $\dfrac{3}{11}$

6 $\dfrac{1}{3}$

7 $\dfrac{1}{9}$

8 $\dfrac{2}{3}$

9 $\dfrac{2}{3}$

10 $\dfrac{4}{5}$

11 $\dfrac{1}{5}$

12 $\dfrac{2}{5}$

13 $\dfrac{7}{12}$

14 $\dfrac{5}{6}$

15 $\dfrac{9}{11}$

16 $\dfrac{4}{5}$

17 $\dfrac{4}{9}$

18 $\dfrac{1}{2}$

19 $\dfrac{1}{9}$

20 $\dfrac{3}{4}$

21 $\dfrac{5}{6}$

22 $\dfrac{8}{9}$

23 $\dfrac{3}{7}$

24 $\dfrac{10}{11}$

25 $\dfrac{3}{5}$

Comparing Fractions

1 3/8 of an apple pie **>** 2/7 of a cherry pie

2 3/9 of a lemon meringue pie **<** 1/2 of a pecan pie

3 4/5 of a pumpkin pie **>** 2/6 of a blueberry pie

4 5/6 of a chocolate crème pie **>** 2/4 of a berry crumble

5 6/9 of a walnut cake **=** 2/3 of the vanilla cupcakes

6 6/7 of the lemon squares **>** 5/6 of the pineapple upside-down cake

7 4/12 of the zucchini bread **=** 1/3 of the cinnamon bread

8 2/3 of the chocolate chip cookies **>** 5/9 of the peanut butter balls

9 6/8 of the chocolate fudge **<** 7/9 of the popcorn balls

10 1/3 of the blueberry muffins **>** 1/4 of the corn muffins

Equivalent Fractions

1) $\dfrac{1}{2} = \dfrac{3}{6} = \dfrac{5}{10} = \dfrac{2}{4} = \dfrac{7}{14} = \dfrac{4}{8} = \dfrac{6}{12}$

2) $\dfrac{28}{32} = \dfrac{49}{56} = \dfrac{21}{24} = \dfrac{42}{48} = \dfrac{35}{40} = \dfrac{14}{18} = \dfrac{7}{8}$

3) $\dfrac{20}{24} = \dfrac{5}{6} = \dfrac{15}{18} = \dfrac{25}{30} = \dfrac{35}{42} = \dfrac{10}{12} = \dfrac{30}{36}$

4) $\dfrac{2}{3} = \dfrac{14}{21} = \dfrac{6}{9} = \dfrac{12}{18} = \dfrac{4}{6} = \dfrac{8}{12} = \dfrac{10}{15}$

5) $\dfrac{4}{16} = \dfrac{2}{8} = \dfrac{7}{28} = \dfrac{6}{24} = \dfrac{3}{12} = \dfrac{1}{4} = \dfrac{5}{20}$

6) $\dfrac{14}{18} = \dfrac{49}{63} = \dfrac{35}{45} = \dfrac{42}{54} = \dfrac{7}{9} = \dfrac{28}{36} = \dfrac{21}{27}$

7) $\dfrac{3}{15} = \dfrac{2}{10} = \dfrac{7}{35} = \dfrac{4}{20} = \dfrac{6}{30} = \dfrac{5}{25} = \dfrac{1}{5}$

8) $\dfrac{63}{70} = \dfrac{9}{10} = \dfrac{27}{30} = \dfrac{36}{40} = \dfrac{45}{50} = \dfrac{54}{60} = \dfrac{18}{20}$

9) $\dfrac{9}{10} = \dfrac{54}{60} = \dfrac{63}{70} = \dfrac{36}{40} = \dfrac{45}{50} = \dfrac{27}{30} = \dfrac{18}{20}$

10) $\dfrac{6}{18} = \dfrac{4}{12} = \dfrac{3}{9} = \dfrac{7}{21} = \dfrac{1}{3} = \dfrac{2}{6} = \dfrac{5}{15}$

Adding Fractions

1. $\dfrac{1}{5} + \dfrac{1}{3} = \dfrac{3}{15} + \dfrac{5}{15} = \dfrac{8}{15}$

2. $\dfrac{1}{6} + \dfrac{2}{3} = \dfrac{1}{6} + \dfrac{4}{6} = \dfrac{5}{6}$

3. $\dfrac{7}{8} + \dfrac{3}{4} = \dfrac{7}{8} + \dfrac{6}{8} = \dfrac{13}{8} = 1\dfrac{5}{8}$

4. $\dfrac{2}{4} + \dfrac{1}{3} = \dfrac{6}{12} + \dfrac{4}{12} = \dfrac{10}{12} = \dfrac{5}{6}$

5. $\dfrac{13}{18} + \dfrac{5}{54} = \dfrac{39}{54} + \dfrac{5}{54} = \dfrac{44}{54} = \dfrac{22}{27}$

6. $\dfrac{3}{4} + \dfrac{3}{10} = \dfrac{15}{20} + \dfrac{6}{20} = \dfrac{21}{20} = 1\dfrac{1}{20}$

7. $\dfrac{1}{2} + \dfrac{3}{5} = \dfrac{5}{10} + \dfrac{6}{10} = \dfrac{11}{10} = 1\dfrac{1}{10}$

8. $\dfrac{1}{3} + \dfrac{1}{4} = \dfrac{4}{12} + \dfrac{3}{12} = \dfrac{7}{12}$

9. $\dfrac{1}{5} + \dfrac{3}{4} = \dfrac{4}{20} + \dfrac{15}{20} = \dfrac{19}{20}$

10. $\dfrac{2}{3} + \dfrac{2}{16} = \dfrac{32}{48} + \dfrac{6}{48} = \dfrac{38}{48} = \dfrac{19}{24}$

11. $\dfrac{1}{4} + \dfrac{5}{9} = \dfrac{9}{36} + \dfrac{20}{36} = \dfrac{29}{36}$

12. $\dfrac{2}{5} + \dfrac{1}{4} = \dfrac{8}{20} + \dfrac{5}{20} = \dfrac{13}{20}$

13. $\dfrac{1}{5} + \dfrac{2}{3} = \dfrac{3}{15} + \dfrac{10}{15} = \dfrac{13}{15}$

14. $\dfrac{1}{4} + \dfrac{18}{46} = \dfrac{23}{92} + \dfrac{36}{92} = \dfrac{59}{92}$

15. $\dfrac{3}{10} + \dfrac{9}{15} = \dfrac{9}{30} + \dfrac{18}{30} = \dfrac{27}{30} = \dfrac{9}{10}$

Dividing Mixed Numbers

1 $3\dfrac{5}{9} \div 3\dfrac{3}{7} = \dfrac{32}{9} \times \dfrac{7}{24} = \dfrac{224}{216} = \dfrac{28}{27} = 1\dfrac{1}{27}$

2 $2\dfrac{1}{3} \div 2\dfrac{1}{4} = \dfrac{7}{3} \times \dfrac{4}{9} = \dfrac{28}{27} = 1\dfrac{1}{27}$

3 $3\dfrac{5}{6} \div 4\dfrac{2}{5} = \dfrac{23}{6} \times \dfrac{5}{22} = \dfrac{115}{132}$

4 $2\dfrac{1}{9} \div 3\dfrac{1}{7} = \dfrac{19}{9} \times \dfrac{7}{22} = \dfrac{133}{198}$

5 $3\dfrac{1}{8} \div 2\dfrac{3}{5} = \dfrac{25}{8} \times \dfrac{5}{13} = \dfrac{125}{104} = 1\dfrac{21}{104}$

6 $3\dfrac{1}{3} \div 3\dfrac{3}{4} = \dfrac{10}{3} \times \dfrac{4}{15} = \dfrac{40}{45} = \dfrac{8}{9}$

7 $2\dfrac{1}{2} \div 2\dfrac{3}{7} = \dfrac{5}{2} \times \dfrac{7}{17} = \dfrac{35}{34} = 1\dfrac{1}{34}$

8 $2\dfrac{7}{9} \div 3\dfrac{3}{7} = \dfrac{25}{9} \times \dfrac{7}{24} = \dfrac{175}{216}$

9 $3\dfrac{1}{2} \div 4\dfrac{1}{3} = \dfrac{7}{2} \times \dfrac{3}{13} = \dfrac{21}{26}$

10 $2\dfrac{5}{8} \div 2\dfrac{5}{7} = \dfrac{21}{8} \times \dfrac{7}{19} = \dfrac{147}{152}$

11 $5\dfrac{3}{8} \div 4\dfrac{1}{3} = \dfrac{43}{8} \times \dfrac{3}{13} = \dfrac{129}{104} = 1\dfrac{25}{104}$

12 $3\dfrac{5}{8} \div 3\dfrac{1}{6} = \dfrac{29}{8} \times \dfrac{6}{19} = \dfrac{174}{152} = \dfrac{87}{76} = 1\dfrac{11}{76}$

13 $4\dfrac{2}{5} \div 2\dfrac{6}{7} = \dfrac{22}{5} \times \dfrac{7}{20} = \dfrac{154}{100} = \dfrac{77}{50} = 1\dfrac{27}{50}$

14 $3\dfrac{2}{7} \div 4\dfrac{3}{5} = \dfrac{23}{7} \times \dfrac{5}{23} = \dfrac{115}{161} = \dfrac{5}{7}$

15 $4\dfrac{1}{2} \div 3\dfrac{2}{3} = \dfrac{9}{2} \times \dfrac{3}{11} = \dfrac{27}{22} = 1\dfrac{5}{22}$

Improper Fractions and Mixed Numbers

(1) $6\frac{1}{2}$

(2) $6\frac{1}{2}$

(3) $2\frac{4}{9}$

(4) $2\frac{4}{5}$

(5) $7\frac{1}{5}$

(6) $6\frac{1}{6}$

(7) $6\frac{5}{8}$

(8) $6\frac{2}{3}$

(9) $5\frac{1}{2}$

(10) $2\frac{9}{10}$

(11) $\frac{19}{4}$

(12) $\frac{11}{3}$

(13) $\frac{24}{5}$

(14) $\frac{29}{5}$

(15) $\frac{13}{2}$

(16) $\frac{40}{7}$

(17) $\frac{29}{4}$

(18) $\frac{10}{3}$

(19) $\frac{37}{5}$

(20) $\frac{55}{6}$

Comparing Decimals

1

0.3, 0.60, 0.65

2

0.546, 0.60, 0.964

3

0.14, 0.23, 0.607

4

0.3, 0.63. 0.76

5

0.459, 0.54, 0.860

6

0.02, 0.11, 0.482

7

0.2, 0.40, 0.5

8

0.71, 0.725, 0.881

9

0.405, 0.50, 0.6

10

0.034, 0.7, 0.9

11

0.39, 0.67, 0.91, 0.92

12

0.513, 0.514, 0.689, 0.796

13

0.305, 0.342, 0.421, 0.422

14

0.898, 0.98, 0.984, 0.989

15

0.843, 0.858, 0.908, 0.921

Adding Decimals

1. $97.88 + 19.37 = 117.25$

2. $69.62 + 75.88 = 145.50$

3. $85.57 + 55.92 = 141.49$

4. $69.54 + 67.99 = 137.53$

5. $57.63 + 43.68 = 101.31$

6. $66.32 + 47.99 = 144.31$

7. $32.67 + 18.87 = 51.54$

8. $21.75 + 48.46 = 70.21$

9. $58.22 + 95.89 = 154.11$

10. $88.77 + 53.96 = 142.73$

11. $33.48 + 68.88 = 102.36$

12. $47.949 + 26.679 = 74.628$

13. $42.9947 + 80.194 = 123.1887$

14. $33.683 + 33.357 = 67.040$

15. $86.416 + 32.333 = 118.749$

Subtracting Decimals from Whole Numbers

1) $2 - 0.55 = 1.45$

2) $4 - 2.64 = 1.36$

3) $8 - 0.36 = 7.64$

4) $80 - 15.3 = 64.7$

5) $246 - 6.4 = 239.6$

6) $377 - 6.3 = 370.7$

7) $460 - 3.6 = 456.4$

8) $943 - 942.1754 = 0.8246$

9) $9 - 6.302 = 2.698$

10) $5323 - 731.2 = 4591.8$

11) $18 - 3.6949 = 14.3051$

12) $38 - 18.2091 = 19.7909$

13) $571 - 4.61 = 566.39$

14) $5 - 1.5011 = 3.4989$

15) $70 - 6.662 = 63.338$

Multiplying Decimals

1. 4.42 x 5.56 = 24.5752

2. 6.23 x 4.37 = 27.2251

3. 2.79 x 4.13 = 11.5227

4. 2.68 x 4.99 = 13.3732

5. 6.82 x 3.48 = 23.7336

6. 7.19 x 7.14 = 51.3366

7. 4.93 x 8.77 = 43.2361

8. 9.22 x 7.72 = 71.1784

9. 3.66 x 2.69 = 9.8454

10. 2.54 x 3.38 = 8.5852

11. 16.23 x 13.49 = 218.9427

12. 80.79 x 10.25 = 828.0975

13. 36.42 x 79.62 = 2899.7604

14. 29.52 x 29.98 = 885.0096

15. 53.18 x 11.27 = 599.3386

Dividing With Decimals

1 $3.60 \div 9 = 0.40$

2 $6.3 \div 3 = 2.1$

3 $6.56 \div 8 = 0.82$

4 $2.76 \div 2 = 1.38$

5 $9.96 \div 4 = 2.49$

6 $4.14 \div 3 = 1.38$

7 $7.75 \div 5 = 1.55$

8 $5.60 \div 5 = 1.12$

9 $5.68 \div 8 = 0.71$

10 $2.76 \div 6 = 0.46$

11 $21.27 \div 3 = 7.09$

12 $17.08 \div 7 = 2.44$

13 $72.36 \div 27 = 2.68$

14 $367.96 \div 4 = 91.99$

15 $938.86 \div 13 = 72.22$

Mixed Operations with Decimals

1   62.454 + 36.748=99.202

2   77.165 + 18.306=95.471

3   44.544 - 22.354=22.19

4   7.18 x 78=560.04

5   12.982 + 15.484=28.466

6   39.638 + 76.042=115.68

7   25.7 x 4.5=115.65

8   9.51 ÷ 6=1.585

9   48.726 - 7.967=40.759

10   55.801 - 49.884=5.917

11   4.67 x 71=331.57

12   5.09 ÷ 8=0.63625

13   81.927 + 62.768=144.695

14   49.7 x 7.9=392.63

15   0.56 ÷ 0.2=2.8

16   8.592 - 5.394=3.198

17   84.882 - 61.501=23.381

18   7.02 ÷ 0.9=7.8

19   0.738 x 0.87=0.64206

20   507.535 ÷ 853=0.595

Converting Decimals to Fractions and Fractions to Decimals

1 $\dfrac{1}{12}$

2 0.6

3 0.75

4 0.667

5 0.833

6 0.6

7 0.091

8 0.875

9 0.444

10 0.944

11 0.769

12 0.571

13 0.053

14 0.36

15 0.286

16 $\dfrac{3}{10}$

17 $\dfrac{27}{50}$

18 $\dfrac{1}{5}$

19 $\dfrac{11}{25}$

20 $\dfrac{1}{100}$

21 $\dfrac{49}{50}$

22 $\dfrac{16}{25}$

23 $\dfrac{37}{100}$

24 $\dfrac{1}{8}$

25 $\dfrac{1}{1}$

Complete the Pattern

1

2

3

4

5

6

7

8

9

10

Numerical Series

1 Alternating subtracting 5 and then adding 10: 20, 30, 25"

2 Odd numbers: 9, 11, 13

3 Multiples of 4: 16, 24, 32

4 Multiplying the previous number by 2: 64, 128, 254

5 Subtracting 6: 55, 49, 43

6 Adding 11: 58, 69, 80

7 Counting down by multiples of 9: 63, 54, 45

8 Add the two previous terms to get the next one: 21, 34, 55

9 Subtracting 3: 34, 31, 28

10 Multiply the previous term by 3: 162, 486, 1458

11 Alternating adding 3 and then subtracting 5 from the previous number: 3, 1, -1

12 Subtract 7: 42, 21, 14

13 Add one more than was added to get the previous term (add 1, then add 2, then add 3): 21, 34, 51

14 Perfect square of the number of the term in the series: 25, 49, 81

15 Subtract 8: -10, -18, -26

16 Alternating adding 2 and then adding 10: 38, 50, 52

17 Alternating adding 6 and then subtracting 5: 24, 30, 25

18 Dividing by 2: 40, 20, 2.5

19 Alternating multiplying by 2 and then adding 2: 22, 44, 92

20 The difference between each term increases by 2 each time (first it's 1, then 3, then 5, then 7...): 27, 38, 51

21 Alternating multiplying the previous term by 3 then adding 3 to the previous term: 174, 522, 525

22 Alternating multiplying the previous term by 2 then subtracting 5 to the previous term: -6, -11, -22

23 Alternating multiplying by 2 then -2: 32, -64, 128

24 Alternating adding 7 then subtracting 5: 20, 27, 22, 29

25 Alternating multiplying by 2 then adding 2: 30, 60, 62

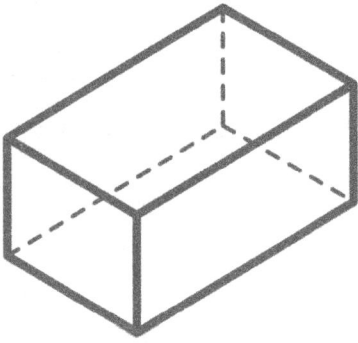

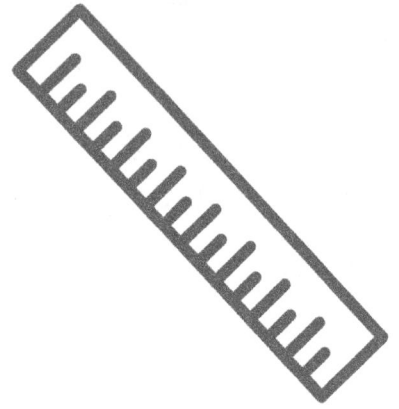

Measurement and Data

Reading Measurements in Two Units

1 4 inches, 10 centimeters

2 10 inches, 25.5 centimeters

3 5 inches, 13 centimeters

4 1 inch, 2.5 centimeters

5 7 inches, 18 centimeters

6 3 inches, 7.5 centimeters

7 12 inches, 30.5 centimeters

8 6 inches, 15.5 centimeters

9 11 inches, 28 centimeters

10 8 inches, 20.5 centimeters

11 2 inches, 5 centimeters

12 9 inches, 23 centimeters

13 3 inches

14 11 centimeters

15 B

Getting Familiar With Solid Figures

1 Triangular Prism: 5 Faces, 9 Edges, 6 Vertices

2 Rectangular prism: 6 Faces, 12 Edges, 8 Vertices

3 Sphere: 0 Faces, 0 Edges, 0 Vertices

4 Square pyramid: 5 Faces, 8 Edges, 5 Vertices

5 Rectangular pyramid: 5 Faces, 8 Edges, 5 Vertices

6 Pentagonal prism: 7 Faces, 15 Edges, 10 Vertices

7 Triangular pyramid: 4 Faces, 6 Edges, 4 Vertices

8 Cylinder: 2 Faces, 0 Edges, 0 Vertices

9 Hexagonal pyramid: 7 Faces, 12 Edges, 7 Vertices

10 Hexagonal prism: 8 Faces, 18 Edges, 12 Vertices

Calculating the Volume of Rectangular Prisms

1 180 cubic units

2 24 cubic units

3 9 cubic units

4 40 cubic units

5 160 cubic units

6 48 cubic units

7 140 cubic units

8 40 cubic units

9 144 cubic units

10 280 cubic units

11 125 cubic cm.

12 512 cubic cm.

13 If the volume of the cube is 8 cubic centimeters, then s3 is 8. That means that s × s × s = 8, so s = 2 cm.

14 If the volume of the cube is 343 cubic centimeters, then s3 is 343. That means that s × s × s = 343, so s = 7 cm.

15 The missing side length is 8 cm because if the volume is 24 cubic centimeters, then 24 = 3 × 1 × x. Solving for x involves dividing both sides by 3 × 1 (which is 3), so x = 8 cm.

16 The missing side length is 9 cm. because if the volume is 360 cubic centimeters, then 24 = 5 × 8 × x. Solving for x involves dividing both sides by 40 (3 × 1), so x = 9 cm.

17 The missing side length is 7 cm. because if the volume is 42 cubic centimeters, then 42 = 2 × 3 × x. Solving for x involves dividing both sides by 6 (3 × 2), so x = 7 cm.

18 The missing side length is 6 cm. because if the volume is 192 cubic centimeters, then 192 = 4 × 8 × x. Solving for x involves dividing both sides by 32 (4 × 8), so x = 6 cm.

19 The missing side length is 9 cm. because if the volume is 450 cubic centimeters, then 450 = 5 × 10 × x. Solving for x involves dividing both sides by 50 (5 × 10), so x = 9 cm.

20 If the volume of the cube is 64 cubic centimeters, then s3 is 64. That means that s × s × s = 64, so s = 4 cm.

Calculating the Volume of Real-World Solids

1 1728 cm³. The equation used to calculate the volume of a cube (a × a × a) or a³. In this problem, the length of a side of the cube is 12 cm, so the volume is calculated by utilizing the formula (12 × 12 × 12) = 1728 cm³.

2 5832 mm³. The equation used to calculate the volume of a cube (a × a × a) or a³. In this problem, the length of a side of the cube is 18 millimeters, so the volume is calculated by utilizing the formula (18 × 18 × 18) = 5832 mm³.

3 231 in³. The equation used to calculate the volume of a rectangular prism, like a cereal box, is length times width times height. Therefore, since our a = 3 in, b = 7 in, and c = 11 in, the volume is calculated by utilizing the formula 3 × 7 × 11 = 231 in³.

4 268 in³. The formula to calculate the volume of a sphere is $\frac{4}{3}\pi r^3$. Therefore, if the radius of a soccer ball is 4 inches, the volume of the sphere is calculated by utilizing the formula $\frac{4}{3}\pi (4)^3 = \frac{4}{3}(64)\pi = 85.3\pi = 268$ in³.

5 600 yd³. The equation used to calculate the volume of a rectangular prism is length times width times height. Therefore, since our l = 25 yd, w = 12 yd, and d = 2 yd, the volume is calculated by utilizing the formula 25 × 12 × 2 = 600 yd³.

6 2,574,466.7 m³. The formula to calculate a pyramid's volume is (L × W × H) ÷ 3. The Great Pyramid has a square base, so it's (230 m × 230 m × 146 m) ÷ 3 = 2,574,466.7 m³. That's huge!

7 21 in³. The formula used to calculate the volume of a cone is $(\frac{1}{3})\pi r^2 h$. Therefore, the volume of the snow cone is calculated by utilizing the formula $\frac{1}{3}\pi 2^2 \times 5 = 6.667\pi$ in³. After substituting 3.14 for π, the volume is 21 in³.

8 288 in³. The equation used to calculate the volume of a rectangular prism, like a chocolate box, is length times width times height. Therefore, since our l = 12 in, w = 8 in, and d = 3 in, the volume is calculated by utilizing the formula 12 × 8 × 3 = 288 in³.

9 150 in³. The equation used to calculate the volume of a rectangular pyramid is $V = \frac{l \times w \times h}{3}$, so plugging in our values gives the answer: $V = \frac{5 \times 9 \times 10}{3} = 150$ in³.

10 36,172.8 m³. The equation used to calculate the volume of a cylinder is $V = \pi r^2 h$, so plugging in our values gives the answer: $V = \pi(12)^2 \times 80 = 11520\pi$ m³. After substituting 3.14 for π, the volume is 36,172.8 m³.

11 840 in³. The equation used to calculate the volume of a rectangular prism is length times width times height. Therefore, since our l = 15 in, w = 7 in, and d = 8 in, the volume is calculated by utilizing the formula 15 × 7 × 8 = 840 in³.

Calculating the Volume of Real-World Solids

12 6912 in³. The equation used to calculate the volume of a rectangular prism is length times width times height. Therefore, since our l = 24 in, w = 18 in, and d = 16 in, the volume is calculated by utilizing the formula 24 × 18 × 16 = 6912 in³.

13 100.48 ft³. The formula used to calculate the volume of a cone is $\frac{1}{3}\pi r^2$ h. Therefore, the volume of the teepee is calculated by utilizing the formula $\frac{1}{3}\pi(4)^2 \times 6 = 96\pi$ ft³. After substituting 3.14 for π, the volume is 100.48 ft³. Keep in mind that the diameter was given (8 feet), so the radius is 4 feet.

14 401.92 m³. The equation used to calculate the volume of a cylinder is $V = \pi r^2$ h, so plugging in our values gives the answer: $V = \pi(2)^2 8 = 32\pi$ m³. After substituting 3.14 for π, the volume is 100.48 m³.

15 350 in³. The equation used to calculate the volume of a rectangular prism is length times width times height. Therefore, since our l = 10 in, w = 7 in, and d = 5 in, the volume is calculated by utilizing the formula 10 × 7 × 5 = 350 in³.

16 50.24 in³. The equation used to calculate the volume of a cylinder is $V=\pi r^2$ h, so plugging in our values gives the answer: $V = \pi(2)^2 4 = 16\pi$ in³. After substituting 3.14 for π, the volume is 50.24 in³.

17 523.3 in³. The formula to calculate the volume of a sphere is $\frac{4}{3}\pi r^3$. Therefore, if the radius of a is 4 inches, the volume of the sphere is calculated by utilizing the formula $\frac{4}{3}\pi(5)^3 = \frac{4}{3}(125)\pi = 166.67\pi = 523.3$ in³.

18 113.04 cm³. The formula to calculate the volume of a sphere is $\frac{4}{3}\pi r^3$. Therefore, if the radius of is 3 cms (half the 6 cm. diameter), the volume of the sphere is calculated by utilizing the formula $\frac{4}{3}\pi(3)^3 = \frac{4}{3}(27)\pi = 36\pi = 113.04$ cm³.

19 1356.48 in³. The formula used to calculate the volume of a cone is $(\frac{1}{3})\pi r^2$ h. Therefore, the volume of the teepee is calculated by utilizing the formula $\frac{1}{3}\pi(6)^2 \times 36 = 432\pi$ in³. After substituting 3.14 for π, the volume is 1356.48 in³. Keep in mind that the diameter was given (12 in), so the radius is 6 in.

20 4.2 in³. The formula to calculate the volume of a sphere is $\frac{4}{3}\pi r^3$. Therefore, if the radius of is 1 inch the volume of the sphere is calculated by utilizing the formula $\frac{4}{3}\pi(1)^3 = \frac{4}{3}\pi = 36\pi = 4.2$ n³.

Weight Word Problems

1 First, we divide 24 hours by the 3-hour increments. 24/3 = 8, so the baby drinks 8 times during the 24 hours. Since the baby has 4 ounces of milk per feeding, we multiply 4 ounces x 8 feedings = 32 ounces.

2 14.4/40 = 0.36 kg per orange.

3 1785 grams = 42.5 kg + 1.785 kg = 44.285 kg.

4 50 pounds – 43 pounds 6 ounces = 6 pounds 10 ounces.

5 First, we take 4 pounds 9 ounces and multiply by three. 4 pound x 3 = 12 pounds, and 9 oz x 3 = 27 oz = 1 pound, 11 ounces. Three end tables are equal to 13 pounds 11 ounces. To subtract 13 pounds 11 ounces form 18 pound 7 ounces, it helps to turn 1 pound of the 18 into its equivalent in ounces(16): 18 pounds 7 ounces – 13 pounds 11 ounces = 17 pounds 23 ounces – 13 pounds 11 ounces. Then we can subtract: 4 pounds 12 ounces.

6 30 grams = 0.3 kg. 6.5 kg / 0.3 kg = 21.67. Therefore, 21 candle molds can be filled.

7 20 pounds 14 ounces + 14 pounds 5 ounces + 15 pounds 8 ounces = 49 pounds 27 ounces = 50 pounds 11 ounces.

8 15 pounds 5 ounces x 3 = 45 pounds 15 ounces. 9 pounds 9 ounces x 3 = 18 pounds 27 ounces or 19 pounds 11 ounces. Then: 45 pounds 15 ounces + 19 pounds 11 ounces = 64 pounds 26 ounces of 65 pounds 10 ounces.

9 (2 x 48.6) + 54.3 + (3 x 84.8 grams) + 103.9 + (2 x 98.4) = 706.6 grams. 894 – 706.6 = 187.4 grams.

10 We need to compare the two ratios here. 8 ounces of butter in 12 cupcakes is 9 / 12 or 0.75 ounces of butter per cupcake. 2 pounds of butter is 2 x 16 = 32 ounces. 32 ounces of butter in 46 cookies is 32/46 = 0.7 ounces of butter per cookie, so the cupcakes have more butter.

11 13 kg = 13000 grams. 13000/450 = 28.9 bowls.

12 96 x 15 = 1440. 2000 - 1440 = 560 pounds. 560/9=5.8. Since you can't have a fraction of a refrigerator, the truck can take 5 more refrigerators.

13 9 pounds 1 ounce - 7 pounds 2 ounces = 1 pound 15 ounces. That is the weight that puppy gains per week so then that weight is added to the current weight: 9 pounds 1 ounce + 1 pound 15 ounces is 11 pounds.

14 9 x 12 = 108 ounces. There are 16 ounces in a pound so 108/16 = 6.75 pounds (6 pounds 12 ounces).

15 5 pounds x 16 ounces per pound is 80 ounces of meat. 80/15 burgers = 5.3 ounces per burger.

Length Word Problems

1 (3 x 35) + (2 x 18) = 105 + 36 = 141 inches of ribbon

2 30 feet 5 inches − 26 feet 9 inches = 29 feet 17 inches − 26 feet 9 inches = 3 feet 8 inches

3 18 − 9 feet 9 inches = 8 feet 3 inches

4 4 + 3 + 3 + 5 + 4 + 4 = 23 inches

5 36 − 23 = 13 inches

6 24 x 1 ÷ 3 = 8 inches

7 9 x 2 ÷ 3 = 6 inches

8 2(12 + 12) + 2 (5 + 5) = 48 + 20 = 68 feet

9 5.5 + 6.75 + 3.75 = 16 miles

10 10 x 3 ÷ 4 = 7.5 inches. 10 − 7.5 = 2.5 inches longer

11 (3 x 36) + (25 x 12) = 108 + 300 = 408 inches

12 28 − 19 = 9 miles further, so dad drives 9 miles more each way. If there are five work days and two drives per day (one each way), 9 miles x 10 drives = 90 more miles per week.

13 4 + 10.5 + 5.25 = 19.75 miles

14 48 x 1 ÷ 4 = 12 inches, so she has knitted 12 inches. 48 - 12 = 36 inches left.

15 11 + 15 + 7.5 = 33.5 inches

16 If each sailboat is 4 inches wide, there are three sailboats per foot (12 ÷ 4 = 3). 44 feet x 3 = 132 sailboats.

17 57 + 49 + 54 = 160 feet for Fiona. 63 + 69 + 29 = 161 feet for Freddie.

18 Freddie won by 1 foot.

19 6 + (6 x 1.5) + 9 + (2 x 9) = 6 + 8 + 9 + 9 = 42 feet

20 27 + (27 x 1 ÷ 3) = 27 + 9 = 36 yards

Estimating Measurements

1 C	**2** B	**3** A	**4** B
5 C	**6** C	**7** B	**8** C
9 C	**10** D	**11** A	**12** C
13 B	**14** B	**15** D	**16** D
17 C	**18** D	**19** A	**20** D

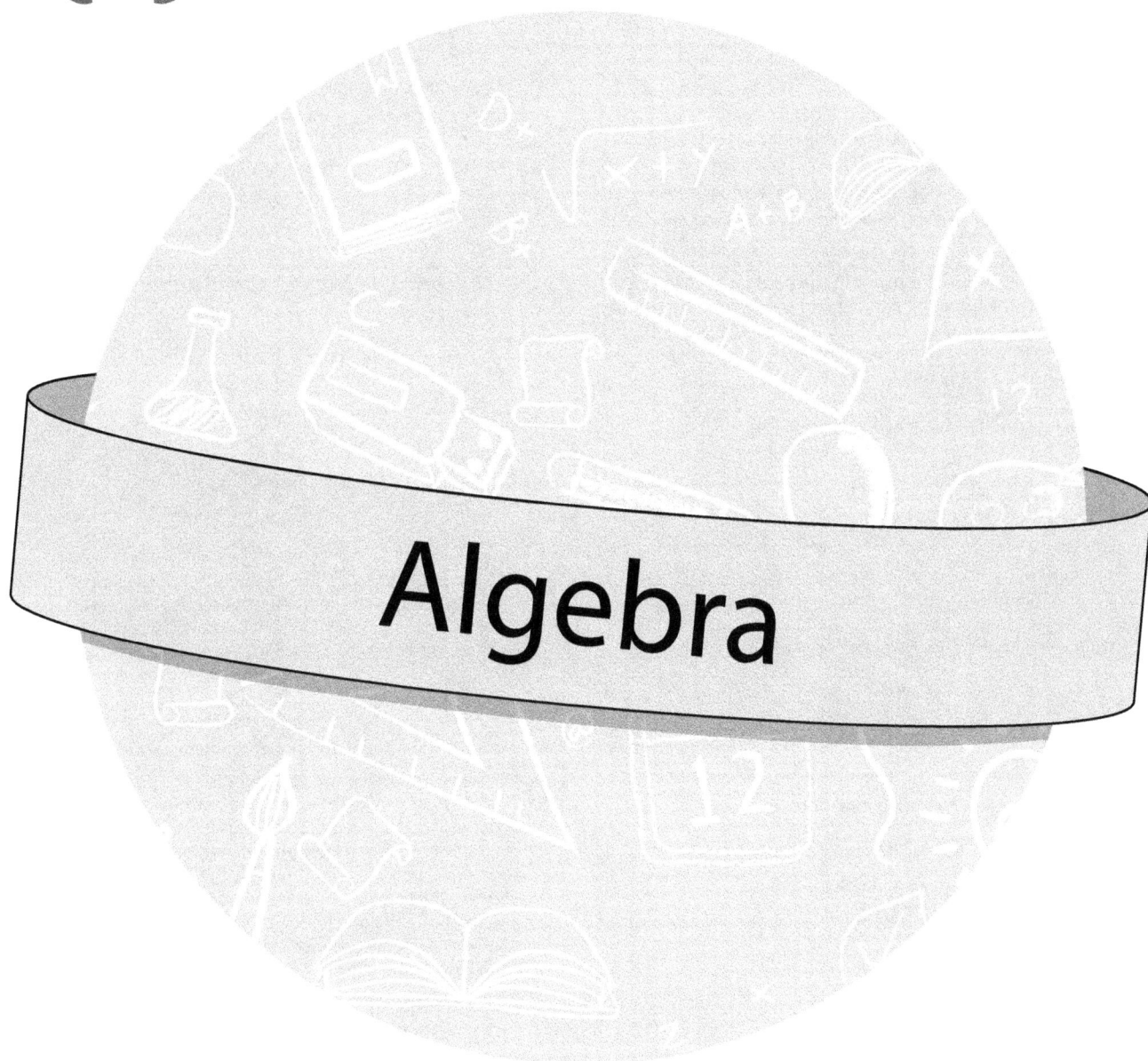

Algebra

Algebraic Word Problems

1 6: This problem involved setting up an algebraic equation to solve for x, or the number of flower trays Carly purchased.
The equation is as follows:
$$6x + 8x = 84$$
So,
$$14x = 84$$
Then divide each side by 14 to solve for x:
$$x = \frac{84}{14} = 6 \text{ trays}$$

2 8: Let a be the number of apples and b the number of bananas. Then, the total cost is 2a + 3b = 22, and it also known that a + b = 10. Using the knowledge of systems of equations, cancel the b variables by multiplying the second equation by -3. This makes the equation -3a - 3b = -30. Adding this to the first equation, the b values cancel to get -a = -8, which simplifies to a = 8.

3 4: Let r be the number of red cans and b be the number of blue cans. One equation is r + b = 10. The total price is $16, and the prices for each can means 1r + 2b = 16. Multiplying the first equation on both sides by -1 results in -r - b = -10. Add this equation to the second equation, leaving b = 6. So, she bought 6 blue cans. From the first equation, this means r = 4; thus, she bought 4 red cans.

4 20 yo-yos: Let y be the number of initial yo-yos in his collection. We can write the following equation from the given information:
$$16 = 6 + \frac{1}{2}y$$
Then we solve for y:
$$10 = \frac{1}{2}y$$
$$y = 20 \text{ yo-yos}$$

5 47: We can represent the pages in the book relative to one another. The lowest page number can be p. Then, the next consecutive page is p + 1, and the third page is p + 2. Therefore, we can write the following equation using only one letter variable:
$$144 = p + (p + 1)+(p + 2)$$
Simplifying and solving yields:
$$144 = 3p + 3$$
$$141 = 3p$$
$$p = 47$$
Thus, the lowest page is 47, then the other two are 48 and 49.

6 43 tickets: Let t be the number of tickets she had before buying the football. Then, we can write the following equation from the given information:
$$26 = \frac{t + 9}{2}$$
Simplifying and solving yields:
$$26 = \frac{t + 9}{2}$$
$$52 = t + 9$$
$$t = 43 \text{ tickets}$$

7 $17: Let c be the cost of each poster. First, we can figure out how much he spent on the posters:
$$\$165 - \$29 = \$136$$
Then, we can write the following equation from the given information:
$$\$136 = 8c$$
$$c = \$17$$

8 46 students: Let s be the number of students per bus. First, we can figure out the number of students who rode a bus:
$$331 - 9 = 322 \text{ students}$$
Then, we can write the following equation from the given information:
$$322 = 7s$$
$$s = 46 \text{ students}$$

9 $9: Let t be the cost of each ticket. Thus, we can write the following equation:
$$13 = \frac{12 + 3t}{2}$$
Simplifying and solving yields:
$$39 = 12 + 3t$$
$$27 = 3t$$
$$t = \$9$$

10 33 cups: First, we have to determine the profit on each cup of lemonade sold. To do this, we start by determining the cost of ingredients per cup. Because each pitcher makes 8 cups, we divide the cost per pitcher by 8: $1.12/8 = $.14 per cup. Then we can determine the profit per cup: $0.75 - $0.14= $0.61 per cup profit. Then, to determine the number of cups she needs to sell: $20.00/$0.61= 32.8. This needs to be rounded up to 33 cups because she can only sell whole cups.

11 74 bottles: This problem requires two equations and two variables. Let's let c be the number of aluminum cans and b be the number of glass bottles. Then we can write the following:
$$29.00 = .05c + .10b$$
And:
$$b = \frac{3}{4}c$$
Therefore, we can substitute this value of b into our first equation to eliminate one of the variables:
$$29.00 = .05c + .10(\frac{3}{4}c)$$
$$29.00 = .125c$$
$$c = 232$$
Therefore, they had 232 cans. To find the number of bottles, we can then use our second equation:
$$b = \frac{3}{4}c$$
$$b = \frac{3}{4}(232)$$
$$b = 174$$

Algebraic Word Problems

12 5 batches: Again, we need two variables and two equations here. Let b be the number of brownie batches and c be the number of cookie batches. Therefore, we can write the following equations from the provided information:

$$184 = 16b + 24c$$
$$c = b + 1$$

Then, we can substitute this value of c relative to b into our first equation to eliminate one variable:

$$184 = 16b + 24c$$
$$184 = 16b + 24(b + 1)$$

Simplifying and solving yields:

$$184 = 16b + 24(b + 1)$$
$$184 = 16b + 24b + 24$$
$$160 = 40b$$
$$b = 4$$

This means that she baked 4 batches of brownies. Since she baked one more batch of cookies than brownies, she baked 5 batches of cookies.

13 33 games: This problem can be solved easily by dividing 44 by 4 (which is 11). This means ¼ of the games, 11 games, are home games. Therefore, the remainder (44 − 11 = 33 games) are away games.

14 9 ladybugs: This is another instance where we can write two equations and use two variables, and then rewrite one variable in terms of the other so that we can solve for one of the variables. We will define l as the number of lady bugs and s as the number of spiders. Because ladybugs are insects, they have six legs, and spiders have eight legs. Therefore, we know the total number of legs (198) is equal to six legs per ladybug times the number of ladybugs plus eight legs per spider times the number of spiders. We also know there are twice as many spiders as ladybugs. Thus, the following equations can be written in this problem:

$$198 = 6l + 8s$$
$$s = 2l$$

Then, we can substitute this value of s relative to l into our first equation to eliminate one variable:

$$198 = 6l + 8s$$
$$198 = 6l + 8(2l)$$

Simplifying and solving yields:

$$198 = 6l + 16l$$
$$198 = 22l$$
$$l = 9$$

Therefore, she counts 9 ladybugs. For completeness, since she counts twice as many spiders as ladybugs, she counts 9 x 2 = 18 spiders. This can be checked by plugging these values into the initial equation:

$$198 = 6(9) + 8(18)$$
$$198 = 54 + 144$$
$$198 = 198$$

15 28 students: This is another relatively simple problem. If there are seven groups with four students each, there are 4 x 7 = 28 students in the class.

16 59.5 minutes: To solve this rate problem, we first need to determine Shankar's pace per mile in his five-mile run. To do this, we divide the total time by 5 miles. 42:30 is equal to 42.5 minutes, since 30 seconds is equal to half of one minute. Thus:

$$\frac{42.5\text{min}}{5 \text{ miles}} = 8.5 \text{ min/mile}$$

Next, we multiply this pace by 7 to find the time it will take him to run 7 miles:

$$8.5 \frac{\text{min}}{\text{mile}} \times 7 \text{ miles} = 59.5 \text{ minutes} = 59 \text{ minutes and } 30 \text{ seconds}$$

Algebraic Word Problems

17 5 days: This is another instance where we can write two equations and use two variables, and then rewrite one variable in terms of the other so that we can solve for one of the variables. We will define l as the number of long days (50 minutes) and s as the number of short days of practice (30 minutes). Because long days are 50 minutes and short days are 30 minutes, we know that the total number of weekly minutes is equal to the number of long days (l) times 50 minutes per l day plus the number of short days (s) times 30 minutes per short day. We also know the sum of the number of l and s days is seven since there are seven days in a week. First, we need to convert the weekly time to minutes so that all times have the same units:

$$5 \text{ hours} \times 60 \tfrac{min}{hour} + 10 \text{ min} = 310 \text{ min}$$

Now we can write our two equations:

$$310 = 50l + 30s$$
$$7 = l + s$$

Next, we can write the second equation in terms of one variable relative to the other:

$$l = 7 - s$$

Now, we can substitute this value of l into our first equation so that we only have one variable to deal with:

$$310 = 50l + 30s$$
$$310 = 50(7 - s) + 30s$$

Simplifying and solving yields:

$$310 = 50(7 - s) + 30s$$
$$310 = 350 - 50s + 30s$$
$$40 = 20s$$
$$s = 2 \text{ days}$$

Therefore, 2 of the days are her shorter 30-minute sessions, which means that 5 days are spent playing 50 minutes since there are 7 days in a week.

18 33 hours: Since we know that her total earnings ($396) is comprised of 1/3 tips and 2/3 of her wages, we can multiply $396 by 2/3 to find the earnings from her wages alone:

$$\$396 \times \frac{2}{3} = \$264$$

Therefore, Sam's mom earned $264 from her wages. Since she makes 8 dollars an hour, we can divide this amount by 8 to find the number of hours worked:

$$\frac{\$264}{\$8/hr} = 33 \text{ hours}$$

19 $324: This problem may seem daunting at first, but we can write an equation with the information we know. She babysat 6 times. Five of those times were four hours and one was six hours. Of the five four-hour times, four were $12 per hour and one was $15 per hour because it had an extra child. Therefore, we can write and solve the following equation:

$$\text{Total earnings} = 4(4 \times \$12) + (4 \times \$15) + (6 \times \$12)$$
$$= 4(\$48) + \$60 + \$72$$
$$= \$192 + \$60 + \$72$$
$$= \$324$$

20 6 boxes: The team needs a total of $270, and each box earns them $3. Therefore, the total number of boxes needed to be sold is 270 ÷ 3, which is 90. With 15 people on the team, the total of 90 can be divided by 15, which equals 6. This means that each member of the team needs to sell 6 boxes for the team to raise enough money to buy new uniforms.

21 30 oranges: One apple/orange pair costs $3 total. Therefore, Jan bought 90 ÷ 3 = 30 total pairs, and hence, she bought 30 oranges.

22 4: Kristen bought four DVDs, which would cost a total of 4 × 15 = $60. She spent a total of $100, so she spent $100 - $60 = $40 on CDs. Since they cost $10 each, she must have purchased 40 ÷ 10 = 4 CDs.

23 390: Three girls for every two boys can be expressed as a ratio: 3:2. This can be visualized as splitting the school into 5 groups: 3 girl groups and 2 boy groups. The number of students that are in each group can be found by dividing the total number of students by 5: 650 divided by 5 equals 1 part, or 130 students per group. To find the total number of girls, the number of students per group (130) is multiplied by how the number of girl groups in the school (3). This equals 390.

24 $62: Kimberley worked 4.5 hours at the rate of $10/h and 1 hour at the rate of $12/h. The problem states that her pay is rounded to the nearest hour, so the 4.5 hours would round up to 5 hours at the rate of $10/h. 5 × $10 + 1 × $12 = $50 + $12 = $62.

25 $0.45: To solve this problem, list the givens:
Store coffee = $1.23/lbs
Local roaster coffee = $1.98/1.5 lbs
Calculate the cost for 5 lbs. of store brand.
$$\frac{\$1.23}{1 \text{ lbs}} \times 5 \text{ lbs} = \$6.15$$
Calculate the cost for 5 lbs. of the local roaster.
$$\frac{\$1.98}{1.5 \text{ lbs}} \times 5 \text{ lbs} = \$6.60$$
Subtract to find the difference in price for 5 lbs.

26 $3,325: List the givens.
1,800 ft. = $2,000
Cost after 1,800 ft.= $1.00/ft.
Find how many feet left after the first 1,800 ft.
3,125 ft.
- 1,800 ft.
1,325 ft.
Calculate the cost for the feet over 1,800 ft.
1,325 ft. × $\frac{\$1.00}{1 \text{ ft}}$ =$1,325
Total for entire cost.
$2,000 + $1,325 = $3,325

27 18: If Ray will be 25 in three years, then he is currently 22. The problem states that Lisa is 13 years younger than Ray, so she must be 9. Sam's age is twice that, which means that the correct answer is 18.

Order of Operations

1
$(10 - 3) \times (9 - 6) + 7^2$
$= (10 - 3) \times (9 - 6) + 49$
$= 7 \times 3 + 49$
$= 21 + 49$
$= 70$

2
$(15 - 7) \times (12 - 6) + 6^2$
$= (15 - 7) \times (12 - 6) + 36$
$= 8 \times 6 + 36$
$= 48 + 36$
$= 84$

3
$(10 - 3)^2 + (12 - 15 \div 5)$
$= 49 + (12 - 3)$
$= 49 + 9$
$= 58$

4
$2 \times (6 \times 3 - 8^2) + 22$
$= 2 \times (6 \times 3 - 8^2) + 22$
$= 2 \times (6 \times 3 - 64) + 22$
$= 2 \times (18 - 64) + 22$
$= 2 \times -46 + 22$
$= -92 + 22$
$= -70$

5
$(9 + 56 - 5) \div 2 + 3^2$
$= (9 + 56 - 5) \div 2 + 3^2$
$= 60 \div 2 + 3^2$
$= 60 \div 2 + 9$
$= 30 + 9$
$= 39$

6
$(11 + 53 - 4^2) \div (9 + 7)$
$= (11 + 53 - 4^2) \div (9 + 7)$
$= (11 + 53 - 16) \div (9 + 7)$
$= (64 - 16) \div (16)$
$= 48 \div 16$
$= 3$

7
$(14 + 19 - 3^2) \div (12 \div 2)$
$= (14 + 19 - 9) \div (12 \div 2)$
$= (33 - 9) \div 6$
$= 24 \div 6$
$= 4$

8
$(3^2 - 4)^2 + (16 + 20 \div 10)$
$= (9 - 4)^2 + (16 + 2)$
$(5)^2 + 18$
$= 25 + 18$
$= 43$

9
$3(8 \div 2^2)^3 + (5 \times 8)$
$= 3(8 \div 4)^3 + 40$
$= 3(2)^3 + 40$
$= 3(8) + 40$
$= 24 + 40$
$= 64$

10
$(2 \times 6^2 + 8) \div (8 - 20 \div 5)^2$
$= (2 \times 36 + 8) \div (8 - 4)^2$
$= 80 \div 16$
$= 5$

11 13: The problem was $4 + (3 \times 2)^2 \div 4$. First, the operation within the parentheses must be completed, yielding: $4 + 6^2 \div 4$. Second, the exponent is evaluated: $4 + 36 \div 4$. Third, the division is conducted: $4 + 9$. Fourth, addition is completed, giving the answer: 13.

12 2: $2 \times (6 + 3) \div (2 + 1)^2$. The first step is to complete the operations in parentheses. $2 \times 9 \div (3)^2$. Then exponents: $2 \times 9 \div 9$. Then multiplication: $18 \div 9$. Then divide: 2.

13 7: For $2^2 \times (3 - 1) \div 2 + 3$, the operations in the parentheses must be completed first: $2^2 \times 2 \div 2 + 3$. Exponents: $4 \times 2 \div 2 + 3$. Multiply: $8 \div 2 + 3$. Divide: $4 + 3$. Add: 7.

Order of Operations

14 65: $(12 + 3) \times (8 - 2) - 5^2$; $15 \times 6 - 5^2$; $15 \times 6 - 25$; $90 - 25 = 65$.

15 114: $(19 - 8) \times (13 - 3) + 2^2$; $11 \times 10 + 2^2$; $11 \times 10 + 4$; $110 + 4 = 114$.

16 48: $(2 + 4)^2 + (9 + 12 \div 4)$; $(6)^2 + (9 + 3)$; $36 + 12 = 48$

17 297: $3 \times (13 \times 3 + 8^2) - 12$; $3 \times (13 \times 3 + 64) - 12$; $3 \times (103) - 12$; $309 - 12 = 297$

18 63: $[(4+3)^2 + 1] + 2^3 - 5$; $(7^2 + 1) + 2^3 - 5$; $50 + 8 - 5 = 53$

19 8: $[6^2 + (20 \div 5 + 4^2)] \div 7$; $[6^2 + (20 \div 5 + 16)] \div 7$; $[6^2 + (4+16)] \div 7$; $(6^2 + 20) \div 7$; $(36+20) \div 7$; $56 \div 7 = 8$

20 -14: $(15/5)^2 - [(12 + 2) + 3^2]$; $(3)^2 - (14 + 3^2)$; $9 - (14 + 9)$; $9 - 23 = -14$

Translating Algebraic Expressions

1 $x + 2$

2 $t - 5$

3 $6(12 + m)$

4 $\dfrac{5}{a}$

5 $b - 4$

6 $\dfrac{2}{5}j - 11h$

7 $9 - \dfrac{3}{4}d$

8 $\dfrac{4}{9s}$ or $4 \div 9s$

9 $\dfrac{2}{3}p + 6$

10 $7g + 4$

11 $\dfrac{1}{2}c + 2d - 5$

12 $4b - \dfrac{1}{3}$

13 $\dfrac{1}{6}(8 + z) - 2x$

14 $\dfrac{8}{(3 + x^2)}$

15 $\dfrac{3}{4}n + \dfrac{1}{3}y + 7$

16 $9k - 5$

17 $4b + 12$

18 $a^2 + 7d$

19 $(3p - 9) \div 2$

20 $2 \times (8 + r)$

21 $\dfrac{3}{5}z \times (y - 1)$

22 $4x - 14y^2$

23 $(\dfrac{1}{5}g + \dfrac{2}{3}m) - 2$

24 $\dfrac{1}{2}g + (10 \div a)$

25 $(d \times (v - 2)) + 15$

Evaluating Algebraic Expressions

1 33: Substitute the given values into the equation:

$7b - 2a = 7(7) - 2(8) = 49 - 16 = 33$.

2 20: The c gets replaced with 4 and the d becomes 8:

$-3 - 9 - 6(4) + 7(8) = -3 - 9 - 24 + 56 = 20$.

3 51: The x gets replaced with 5 and the y becomes 3:

$6x + 7y = 6(5) + 7(3) = 30 + 21 = 51$.

4 24: The m is replaced with 3 and the n becomes 6:

$-2(8m - 6n) = -2(8(3) - 6(6)) = -2(24 - 36) = -2(-12) = 24$.

5 10: Each instance of x is replaced with 2, and each instance of y is replaced with 3 to get $2^2 - 2 \times 2 \times 3 + 2 \times 3^2 = 4 - 12 + 18 = 10$

6 608: Each instance of n is replaced with 4 to get:

$8n + 5n^3 + 16n^2$

$= 8(4) + 5(4)^3 + 16(4)^2$

$= 32 + 320 + 256$

$= 608$

7 -2028: Each instance of t is replaced with -2, and each instance of g is replaced with 7 to get

$(15 - 8t^2) - (5g^3 - 9 + 6g^2) + (3 + 7t)$

$= (15 - 8(-2)^2) - (5(7)^3 - 9 + 6(7)^2) + (3 + 7(-2))$

$= (15 - 32) - (1715 - 9 + 294) + (3-14)$

$= (-17) - (2000) + (-11)$

$= -2028$

8 -180,072: Each instance of t is replaced with 3, and each instance of b is replaced with -6 to get

$(2a^2 + 6a^4 - 4a)(3b^3 + 8b^2 + b)$

$= (2(3)^2 + 6(3)^4 - 4(3))(3(-6)^3 + + 8(-6)^2 + (-6))$

$= (18 + 486 - 12)(-648 + 288 - 6)$

$= (492)(-366)$

$= -180,072$

9 1008: Each instance of k is replaced with 12, and each instance of l is replaced with -8 to get

$9k^2 - 6l^2 + 8k$

$= 9(12)^2 - 6(-8)^2 + 8(12)$

$= 1296 - 384 + 96$

$= 1,008$

Evaluating Algebraic Expressions

10 21: Each instance of b is replaced with 2 to get
$$(8b^2 + 3) + (58 - 2b^3) - (6b^3 + 4b)$$
$$= (8(2)^2 + 3) + (58 - 2(2)^3) - (6(2)^3 + 4(2))$$
$$= (32 + 3) + (58 - 16) - (48 + 8)$$
$$= (35) + (42) - (56)$$
$$= 21$$

11 -460: Each instance of c is replaced with -4, and each instance of d is replaced with 3 to get
$$(8c^3 - 6c^2 + 4) + (3d^3 + 7c^2)$$
$$= (8(-4)^3 - 6(-4)^2 + 4) + (3(3)^3 + 7(3)^2)$$
$$= (-512 - 96 + 4) + (81 + 63)$$
$$= (-604) + 144$$
$$= -460$$

12 -3948: Each instance of x is replaced with -3, and each instance of y is replaced with 7 to get
$$y(9 - 7x^4 + 2x)$$
$$= 7(9 - 7(-3)^4 + 2(-3))$$
$$= 7(9 - 567 - 6)$$
$$= 7(-564)$$
$$= -3,948$$

13 828: Each instance of c is replaced with 15, and each instance of d is replaced with 4 to get
$$(8 + 4c^2) - (2d^3 - 3d^2)$$
$$= (8 + 4(15)^2) - (2(4)^3 - 3(4)^2)$$
$$= (908) - (128-48)$$
$$= 828$$

14 15: Each instance of m is replaced with 1, and each instance of n is replaced with -1 to get
$$(2m^2 + 3)(4n^2 - 7n)$$
$$= (2(1)^2 + 3)(4(-1)^2 - 7(-1))$$
$$= (5)(3)$$
$$= 15$$

Solving Equations

1 -5x = -40. Divide each side by -5. X = 8

2 2 + j = -8. Subtract 2 from both sides, which yields j = -10.

3 -7 + a = -10. Add 7 to both sides, which gives a = -3.

4 -8h + 6h = 22.
Simplify the left side of the equation first: -2h = 22.
Then divide both sides by -2: h = -11.

5 12 = m − 2.
Add 2 to both sides, which gives the value m = 14.

6 11 = c − 3.
Add 3 to both sides, which gives the value of c = 14.

7 -12 = 2y. Divide both sides by 2. Then y = -6.

8 7f = 56. Divide both sides by 7, which yields f = 8.

9 -34 = 6.8c. Divide both sides by 6.8, which yields c = -5.

10 6g = 36. Divide both sides by 6, which yields g = 6.

11 $\frac{z}{4}$ = 7. Multiply both sides by 4 to get rid of the fraction. This leaves z = 28.

12 -5.1d = -35.7. Divide both sides by -5.1: d = 7.

13 $\frac{b}{3}$ = -4. Multiply both sides by 3 to get rid of the fraction. This leaves b = -12.

14 2.5t = 10. Divide both sides by 2.5, which yields t = 4.

15 46.4 = -5.8a. Divide both sides by -5.8, which yields a = -8.

16 $\frac{j}{5}$ = 4.2. Multiply both sides by 5: j = 21.

17 31.8 = 5.3x. Divide both sides by 5.3, which yields x = 6.

18 -16 = -2b -4 + 5b.
Simplify the right side of the equation: -16 = 3b -4.
Then add 4 to both sides and then divide both sides by 3 to isolate the variable. The result is b = -4.

19 9 = -23u − 22. Add 22 to both sides and then divide both sides by -23 to isolate the variable. The result is u = -1.35.

20 23 = 13d + 2. Subtract 2 from both sides and then divide both sides by 13 to isolate the variable. The result is d = 1.62.

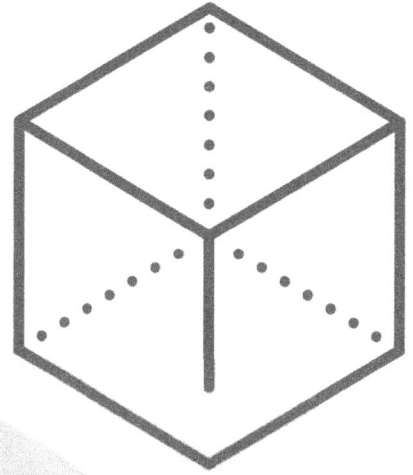

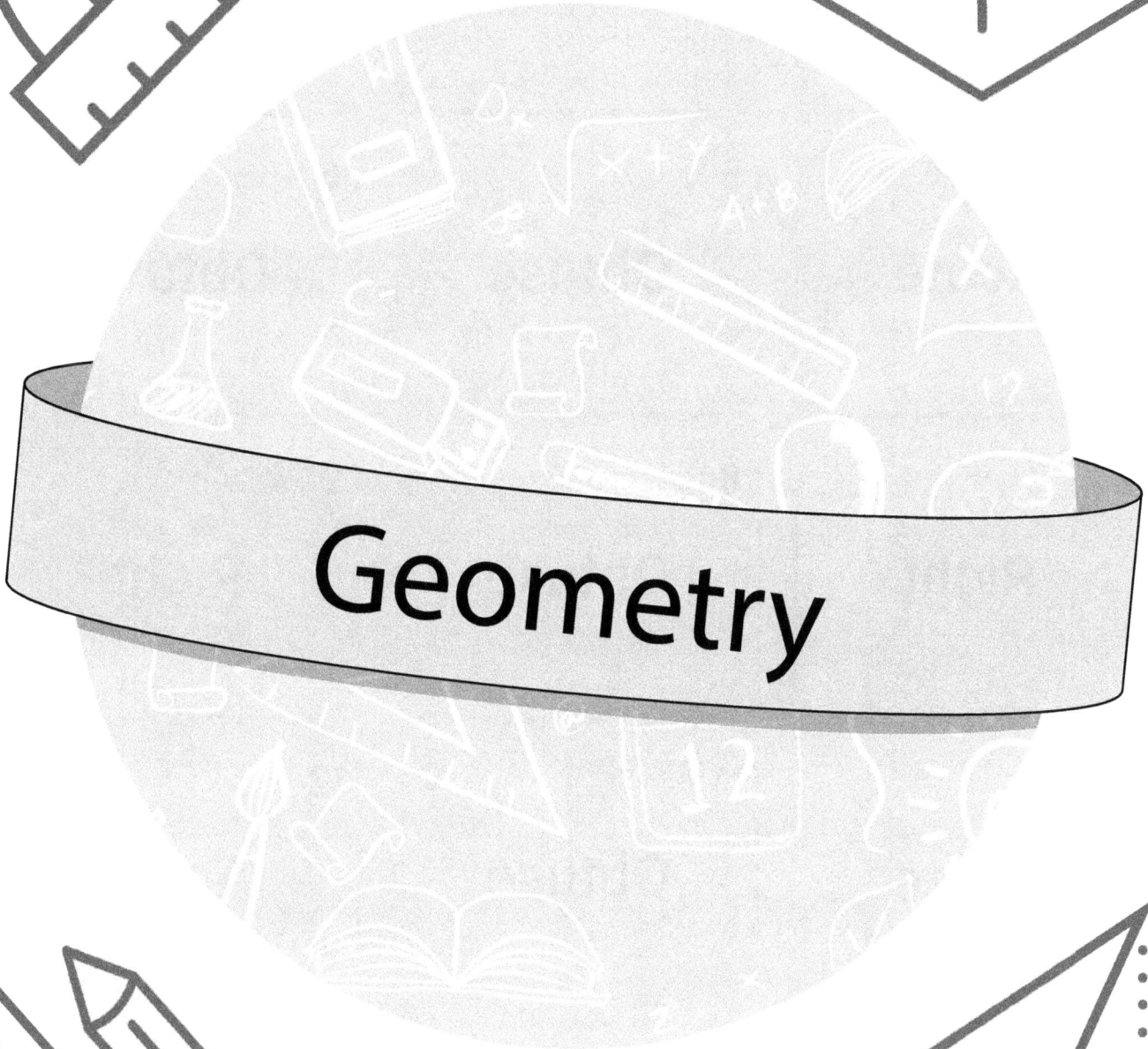

Geometry

Classifying Angles

1 Obtuse

2 Acute

3 Right

4 Acute

5 Obtuse

6 Obtuse

7 Right

8 Obtuse

9 Right

10 Acute

11 Obtuse

12 Acute

13 Acute

14 Right

15 Acute

Determining the Measurement of Missing Angles

1

25°

2

30°

3

60°

4

21°

5

48° + x° = 116°
x = 116 − 48
x = 68°

6

x° = 23° + 11° = 34°

7

35° + x° = 84°
x = 84 − 35
x = 49°

8

x° = 91° + 48° = 139°

9

x° = 27° − 12° = 15°

10

33°

Find the Complementary Angle

1. 80°
2. 35°
3. 60°
4. 10°
5. 78°
6. 63°
7. 52°
8. 19°
9. 27°
10. 34°

Find the Supplementary Angle

1. 135°
2. 26°
3. 141°
4. 13°
5. 161°
6. 90°
7. 44°
8. 107°
9. 97°
10. 64°
11. 110°
12. 52°
13. 93°
14. 1°
15. 61°

Finding Missing Angles in Quadrilaterals

1 95°

2 87°

3 53°

4 113°

5 62°

6 119°

7 109°

8 53°

9 93°

10 59°

11 102

12 101°

13 92°

14 110°

15 87°

16 99°

17 114°

18 107°

19 73°

20 88°

Points, Lines, and Planes

1 U, V, W, X, and Y

2 $\overline{(UV)}$, $\overline{(VY)}$, $\overline{(VX)}$, $\overline{(VW)}$

3 U, V, and W

4 $\overleftrightarrow{(XY)}$ and $\overleftrightarrow{(WU)}$

5 Y

6 $\overrightarrow{(VX)}$, $\overrightarrow{(VW)}$, $\overrightarrow{(VY)}$, $\overrightarrow{(VU)}$

7 N or K

8 $\overline{(NM)}$

9 Q, P

10 $\overleftrightarrow{(LP)}$ and $\overleftrightarrow{(KN)}$

11 M

12 $\overrightarrow{(QK)}$, $\overrightarrow{(QP)}$, $\overrightarrow{(QL)}$, $\overrightarrow{(QN)}$

13 L and T

14 B

15 D or E

16 $\overleftrightarrow{(AC)}$

17 E with either A, B, or C

18 Line n

19 E

Parallel and Intersecting Lines

1 Parallel

2 Parallel

3 Perpendicular

4 Parallel

5 Intersecting

6 Parallel

7 Intersecting

8 Perpendicular

9 Perpendicular

10 Intersecting

Calculating Perimeter

1 32 inches. 8 inches x 4 sides = 32 inches

2 31.4 cm. P = 2πr = 10π cm.

3 The first triangle. The first triangle has a perimeter of 4 + 9 + 8 inches, which is 21 inches, while the perimeter of the equilateral triangle is 3 x 6 inches, which is 18 inches. Therefore, the first triangle has a larger perimeter.

4 42 cm. A hexagon has six sides, so 6 x 7 cm = 42 cm.

5 242 meters. The perimeter of a rectangle is P=2l+2w so in this case, P=2(37)+2(84) = 242 meters.

6 32 cm.

7 60 mm.

8 20 inches.

9 68 yards.

10 14π or 43.96 feet.

11 72 cm.

12 50 inches.

13 42 cm.

14 64 feet.

15 15 yards.

16 64 inches.

17 24 inches.

18 If the length is twice the width, the length is 2 x 8 = 16. 16 + 16 + (2 x 8) = 48 cm.

19 14 + 14 + (14 / 2) = 35 inches.

20 27 / 3 = 9. 9 + 9 + 27 + 27 = 72 cm.

Finding the Missing Side Lengths

1 9 inches. The perimeter of a rectangular paper is 32 inches. One side is 7 inches. That means there are 14 inches for those two sides. 32 – 14 = 18 inches left for the other two sides. 18 inches divided by two more sides is 9 inches per side.

2 9 inches. 36 / 4 = 9 inches.

3 5 inches. 20 / 4 = 5 inches.

4 58 - (2 x 14) = 30 inches. 30 / 2 = 15 inches.

5 15 inches. 36 - 20 = 16. 16 / 2= 8 inches.

6 17 yards. 65 - (29 + 19) = 17 yards.

7 37 feet. 148 / 4 = 37 feet.

8 4 cm. 19 - (7 + 8) = 4 cm.

9 48 / 6 = 8 cm.

10 6 cm. 17 - (5 + 6) = 6 cm.

11 12 inches. 96 / 8 = 12 inches.

12 26 meters. 122 - (2 x 35) = 52. 52 / 2 = 26 meters.

13 27 inches. 90 - (2 x 18) = 54. 52 / 2 = 27 inches.

14 23 inches. 140 - (2 x 47) = 46. 46 / 2 = 23 inches.

15 9 cm. 45 / 5 = 9 cm.

16 12 cm. 42 – (17 + 13) = 12 cm.

17 20 mm. 85 – (17 + 11 + 19 + 18) = 20 mm.

18 30 cm. 120 / 4 = 30 cm.

19 5 cm. 9 + 12 = 21. 26 – 21 = 5 cm.

20 4 inches.
14, 13, 18, 22, 16. 14 + 13 + 18 + 22 + 16 = 83 in, so the remaining side is 4 inches.

Area of Triangles

1 We are given a base (a) and height (b) for this right triangle, so we just need to plug the values into the formula for the area of a triangle to calculate the area: $A = \frac{1}{2}bh = \frac{1}{2}4 \times 3 = 6$ yd².

2 This is a scalene triangle, but we are given the measurement of the height, so we can plug the value of the base (a) and the height into the formula to calculate the area: $A = \frac{1}{2}bh = \frac{1}{2}8.5 \times 3 = 12.75$ cm².

3 This is a scalene triangle, but we are given the measurement of the height, so we can plug the value of the base (a) and the height into the formula to calculate the area:
$A = \frac{1}{2}bh = \frac{1}{2}98 \times 48 = 2352$ ft².

4 We are given a base (a) and height (b) for this right triangle, so we just need to plug the values into the formula for the area of a triangle to calculate the area: $A = \frac{1}{2}bh = \frac{1}{2}5 \times 12 = 30$ ft².

5 Because this is an isosceles triangle, we know that the two equal sides (labeled b) can be taken as the hypotenuse of a right triangle formed if we drop a median line in the triangle bisecting the base, a. Then, the hypotenuse is 65 meters. To calculate the height, we have to apply the Pythagorean Theorem using our bisected base with a resultant length of 20.5 meters. Therefore, we know that $a^2 + b^2 = c^2$, or $65^2 - 20.5^2 = height^2$. So, $4225 - 420.25 = 3804.75$, so the height is 61.6 meters. Therefore, the area $A = \frac{1}{2}bh = \frac{1}{2}65 \times 61.6 = 2002$ m².

6 If each side is 4, the hypotenuse is 4 feet. To calculate the height, we have to apply the Pythagorean Theorem ($a^2 + b^2 = c^2$). Because it is an equilateral triangle, the height will be located along the median line of the triangle, which means the base would be bisected, with each half being 2 feet. Therefore, we know that $4^2 - 2^2 = height^2$, so $16 - 4 = 12$. Therefore, the height is 3.46 feet. Therefore, the area $A = \frac{1}{2}bh = \frac{1}{2}4 \times 3.46 = 6.92$ ft².

7 We are given a base (a) and height (b) for this right triangle, so we just need to plug the values into the formula for the area of a triangle to calculate the area: $A = \frac{1}{2}bh = \frac{1}{2}24 \times 7 = 84$ cm².

8 Because this is an isosceles triangle, we know that the two equal sides (labeled b) can be taken as the hypotenuse of a right triangle formed if we drop a median line in the triangle bisecting the base, a. Then, the hypotenuse is 5 yards. To calculate the height, we have to apply the Pythagorean Theorem using our bisected base with a length of 2 yards. Therefore, we know that $5^2 - 2^2 = height^2$. So, $25 - 4 = 21$; thus, the height is 4.58 yards. Therefore, the area is $A = \frac{1}{2}bh = \frac{1}{2}4 \times 4.58 = 9.16$ yd².

9 We are given a base (a) and height (b) for this right triangle, so we just need to plug the values into the formula for the area of a triangle to calculate the area: $A = \frac{1}{2}bh = \frac{1}{2}12 \times 16 = 96$ mm².

Area of Triangles

10 If each side is 8, the hypotenuse is 8 inches. To calculate the height, we have to apply the Pythagorean Theorem. Because it is an equilateral triangle, the height will be located along the median line of the triangle, which means the base would be bisected, with each half being 4 inches.
Therefore, we know that $8^2 - 4^2 = height^2$. So, $64 - 16 = 48$ inches; thus, the height is 6.93 inches.
Therefore, the area is $A = \frac{1}{2}bh = \frac{1}{2}8 \times 6.93 = 27.72$ in².

11 This is a scalene triangle, but we are given the measurement of the height, so we can plug the value of the base (a) and the height into the formula to calculate the area:
$A = \frac{1}{2}bh = \frac{1}{2}98 \times 48 = 2352$ ft².

12 Because this is an isosceles triangle, we know that the two equal sides (labeled b) can be taken as the hypotenuse of a right triangle formed if we drop a median line in the triangle bisecting the base, a. Then, the hypotenuse is 12 feet. To calculate the height, we have to apply the Pythagorean Theorem using our bisected base with a resultant length of 5 feet.
Therefore, we know that $12^2 - 5^2 = height^2$. So, $144 - 25 = 119$; thus, the height is 10.91 feet.
Therefore, the area is $A = \frac{1}{2}bh = \frac{1}{2}10 \times 10.91 = 54.55$ ft².

13 We are given a base (a) and height (b) for this right triangle, so we just need to plug the values into the formula for the area of a triangle to calculate the area: $A = \frac{1}{2}bh = \frac{1}{2}70 \times 56 = 1960$ yd²

14 We are given a base (a) and height (b) for this right triangle, so we just need to plug the values into the formula for the area of a triangle to calculate the area: $A = \frac{1}{2}bh = \frac{1}{2}75 \times 55 = 2062.5$ mm²

15 Because this is an isosceles triangle, we know that the two equal sides (labeled b) can be taken as the hypotenuse of a right triangle formed if we drop a median line in the triangle bisecting the base, a. Then, the hypotenuse is 72 inches. To calculate the height, we have to apply the Pythagorean Theorem using our bisected base with a resultant length of 29 inches. Therefore, we know that $a^2 + b^2 = c^2$, or $72^2 - 29^2 = height^2$. So, $5184 - 841 = 4343$, so the height is 65.90 inches.
Therefore, the area $A = \frac{1}{2}bh = \frac{1}{2}58 \times 65.90 = 1911$ in.²

Area of Circles

	Exact Circumference	Approximate Circumference
1 A circle with a radius of 2 meters.	4π sq. meters	12.56 sq. meters
2 A circle with a radius of 1 inch.	π sq. inches	3.14 sq. inches
3 A circle with a diameter of 12 feet.	36π sq. feet	113.04 sq. feet
4 A circle with a diameter of 26 mm.	169π sq. millimeters	530.66 sq. mm.
5 A circle with a diameter of 18 feet.	81π sq. feet	254.34 sq. feet
6 A circle with a radius of 10 meters.	100π sq. meters	314 sq. meters
7 A circle with a radius of 5 centimeters.	25π sq. centimeters	78.5 sq. cm.
8 A circle with a radius of 8 inches.	64π sq. inches	200.96 sq. inches
9 A circle with a diameter of 14 inches.	49π sq. inches	153.86 sq. inches
10 A circle with a diameter of 24 feet.	144π sq. feet	452.16 sq. feet

Area of Mixed Shapes and Figures

1) 9 m². The formula for the area of a square is $A = s^2$, so a square with a side length of 3 meters has an area of $A = s^2 = 3^2 = 9$ m².

2) 55 ft². The formula for the area of a rectangle is $A = lw$, so a rectangle with length of 11 feet and a width of 5 feet has an area of $A = lw = 11 \times 5 = 55$ ft².

3) 56 in². The formula for the area of a rectangle is $A = lw$, so a rectangle with length of 8 inches and a width of 7 inches has an area of $A = lw = 8 \times 7 = 56$ in².

4) 36 cm². The formula for the area of a square is $A = s^2$, so a square with a side length of 6 centimeters has an area of $A = s^2 = 6^2 = 36$ cm².

5) 14 in². The area of a triangle is $A = \frac{1}{2} \times b \times h$, so a triangle with a base of 4 in. and a height of 7 in. has an area of $A = \frac{1}{2} \times 4 \times 7 = 14$ in².

6) 48 cm². The area of a triangle is $A = \frac{1}{2} \times b \times h$, so a triangle with a base of 12 cm and a height of 8 cm has an area of $A = \frac{1}{2} \times 12 \times 8 = 48$ cm².

7) 225 in². The formula for the area of a square is $A = s^2$, so a square with a side length of 15 inches has an area of $A = s^2 = 15^2 = 225$ in².

8) 60 in². The formula for the area of a rectangle is $A = lw$, so a rectangle with length of 12 inches and a width of 5 inches has an area of $A = lw = 12 \times 5 = 60$ in².

9) 254.34 cm². The area of a circle is calculated through the formula $A = \pi \times r^2$. Therefore, a circle with a radius of 9 cm has an area of $A = \pi \times r^2 = A = \pi \times 9^2 = 81\pi = (81 \times 3.14) = 254.34$ cm².

10) 10 cm². The area of a triangle is $A = \frac{1}{2} \times b \times h$, so a triangle with a base of 4 cm and a height of 5 cm has an area of $A = \frac{1}{2} \times 4 \times 5 = 10$ cm².

11) 914 yd². This compound figure has a rectangle and two semicircles. Therefore, the area is found by founding the area of a rectangle that is 30 yards by 20 yards and a circle with a radius of 10 yards. Note that the width of the rectangle is the diameter of the circle, so it is 2 x 10 or 20 yards. The area is $A = lw + (\pi \times r^2) = (20 \times 30) + (\pi \times 10^2) = 600 + 100\pi = 914$ yd².

12) 12.3 in². This compound figure has a triangle and a semicircle. Therefore, the area is found by founding the area of a triangle that has a base of 4 inches and a height of 3 inches and half the area of a circle with a radius of 2 inches. Note that the base of the triangle is the diameter of the circle, so it is 2 x 2 or 4 inches.
The area is $A = (\frac{1}{2} \times b \times h) + \frac{1}{2}(\pi \times r^2) = (\frac{1}{2} \times 4 \times 3) + \frac{1}{2}(\pi \times 2^2) = 6 + 2\pi = 12.3$ in².

13) 688.5 cm². This compound figure has a rectangle and a semicircle. Therefore, the area is found by founding the area of a rectangle that is 21 cm. by 28 cm. and half the area of a circle with a radius of 8 cm. The area is $A = lw + \frac{1}{2}(\pi \times r^2) = (21 \times 28) + \frac{1}{2}(\pi \times 8^2) = 588 + 32\pi = 688.5$ cm².

Area of Mixed Shapes and Figures

14 61.1 ft². This compound figure has two equal triangles and a semicircle. Therefore, the area is found by founding the area of a triangles with a base of 4 feet and a height of 9 feet and doubling it (since there are two) and then adding it to half the area of a circle with a radius of 4 feet. Note that the base of the triangle is the radius of the circle because two triangles fit across the full diameter of the circle.

The area is $A = 2(\frac{1}{2} \times b \times h) + \frac{1}{2}(\pi \times r^2) = 2(\frac{1}{2} \times 9 \times 4) + \frac{1}{2}(\pi \times 4^2) = 36 + 8\pi = 61.1$ ft².

15 370.2 m². This compound figure has a triangle and a semicircle. Therefore, the area is found by founding the area of a triangle that has a base of 18 meters and a height of 27 meters and half the area of a circle with a radius of 9 meters. Note that the base of the triangle is the diameter of the circle, so it is 2 x 9 or 18 meters.

The area is $A = (\frac{1}{2} \times 18 \times 27) + \frac{1}{2}(\pi \times 9^2) = (\frac{1}{2} \times 4 \times 3) + \frac{1}{2}(\pi \times 2^2) = 243 + 40.5\pi = 370.2$ m².

16 354 m². The area of this compound figure can be found by adding the area of two rectangles within it. The first is on the left: 14 m x 15 m. The second is the smaller part on the right, which is actually a square (12 x 12 m). Therefore the area of the figure is the sum of those two components: (14 x 15) + (12 x 12) = 354 m².

17 300 yd². The area of this compound figure can be found by adding the area of two rectangles within it. The first is on the bottom half: 10 yd x 20 yd. The second is the smaller part on the top half, which is actually a square (10 x 10 yd). Therefore the area of the figure is the sum of those two components: (10 x 20) + (10 x 10) = 300 yd².

18 800 cm². The area of this compound figure can be found by adding the area of two rectangles within it. The first is on the bottom half: 10 cm x 40 cm. The second is the smaller part on the top half, which is actually a square (20 x 20 cm). Therefore the area of the figure is the sum of those two components: (10 x 40) + (20 x 20) = 800 cm².

19 33.8 in². This compound figure has a triangle and a rectangle. Therefore, the area is found by founding the area of a triangle that has a base of 4.5 inches and a height of 3 inches and the area of a rectangle with a length of 6 inches and a width of 4.5 inches. Note that the base of the triangle is the width of the rectangle, so it is 4.5 inches.

The area is $A = (\frac{1}{2} \times b \times h) + lw = (\frac{1}{2} \times 4.5 \times 3) + (4.5 \times 6) = 6.75 + 27 = 33.8$ in².

20 262 in². This compound figure has a triangle and a semicircle. Therefore, the area is found by founding the area of a triangle that has a base of 10 inches and a height of 21 inches and half the area of a circle with a radius of 10 inches. Note that the base of the triangle is the radius of the circle, so it is 10 inches.

The area is $A = (\frac{1}{2} \times b \times h) + \frac{1}{2}(\pi \times r^2) = (\frac{1}{2} \times 10 \times 21) + \frac{1}{2}(\pi \times 10^2) = 105 + 50\pi = 262$ in².

Lines of Symmetry in the Alphabet

1 A

2 B

3 C

4 D

5 E

6 No symmetry

7 No symmetry

8 H

9 I

10 No symmetry

11 No symmetry

12 No symmetry

13 M

14 No symmetry

15 O

Lines of Symmetry in the Alphabet

16 No symmetry

17 No symmetry

18 No symmetry

19 No symmetry

20

21

22

23

24

25

26 No symmetry

Reading the Coordinate Plane

1 B (4, 4)

2 C (9, 1)

3 E (0, 9)

4 F (9, 5)

5 G (5, 9)

6 H (8, 3)

7 J (9, 3)

8 L (7, 2)

9 M (2, 4)

10 O (2, 2)

11 Q (2, 1)

12 R (8, 2)

13 S (7, 4)

14 T (0, 3)

15 U (0, 5)

16 V (1, 2)

17 W (9, 4)

18 X (3, 5)

Reading the Coordinate Plane

1 A (9, 0)

2 C (0, 3)

3 E (2, 2)

4 F (6, 9)

5 G (8, 3)

6 H (0, 4)

7 I (3, 2)

8 K (9, 4)

9 L (7, 0)

10 M (7, 4)

11 N (7, 3)

12 Q (8, 4)

13 R (4, 5)

14 S (8, 6)

15 T (0, 2)

16 U (3, 6)

17 V (2, 1)

18 W (1, 6)

Plotting Points in the Coordinate Plane

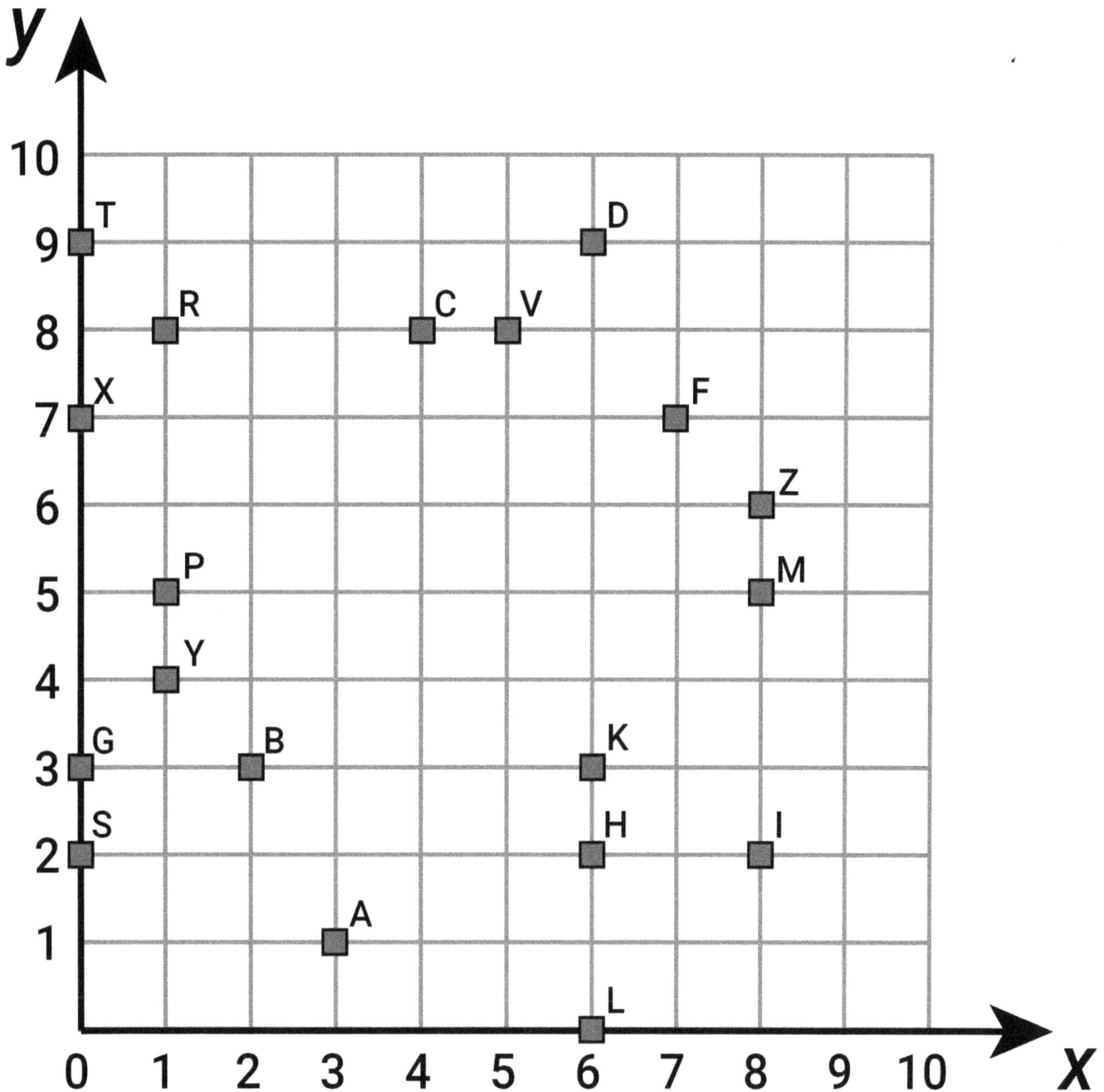

Plotting Points in the Coordinate Plane

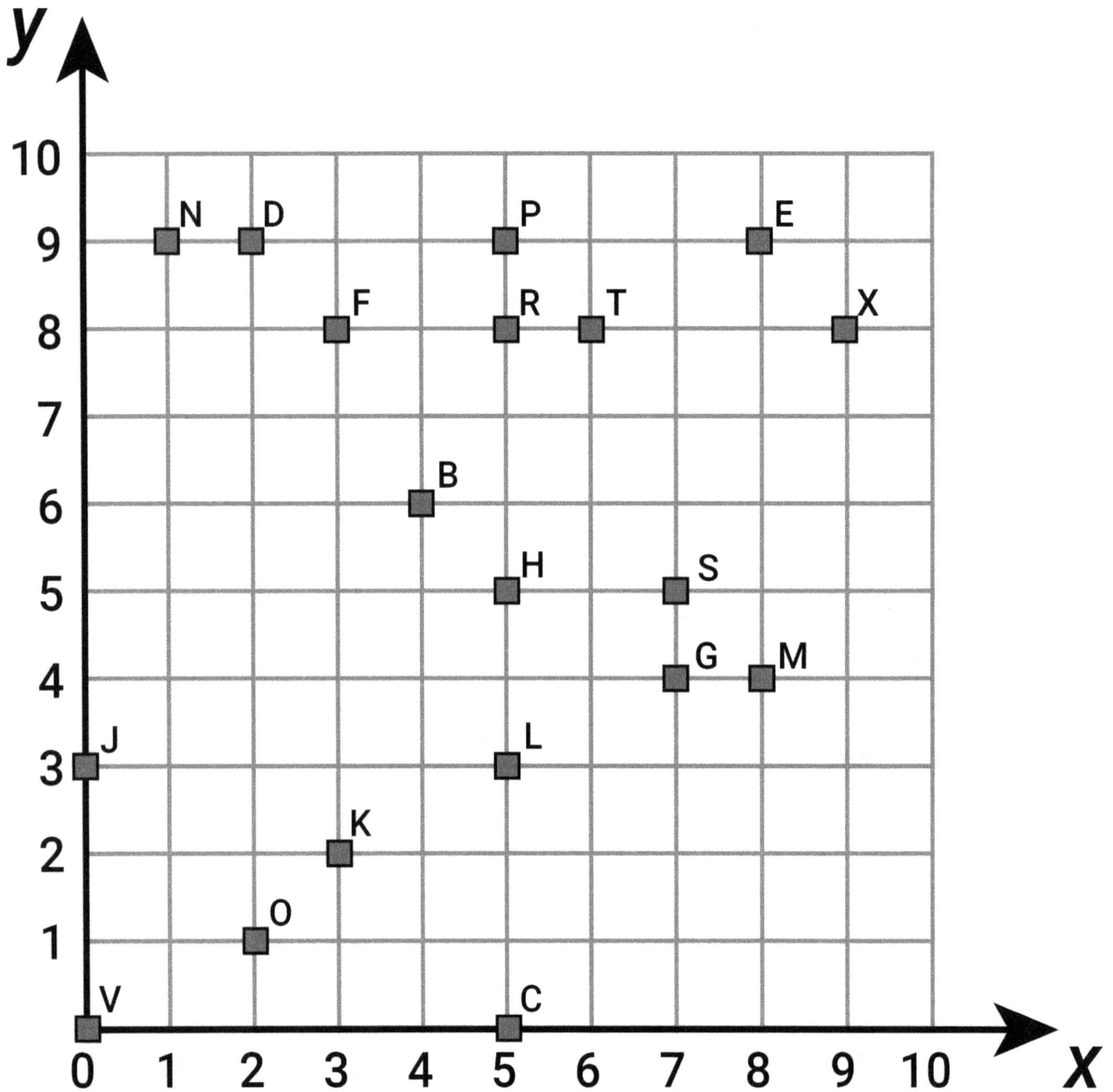

Identifying Shapes

1. Circle
2. Triangle
3. Square
4. Septagon/heptagon
5. Circle
6. Rectangle
7. Rectangle
8. Triangle
9. Triangle
10. Pentagon
11. Hexagon
12. Septagon/heptagon
13. Pentagon
14. Rectangle
15. Octagon
16. Circle
17. Pentagon
18. Rectangle
19. Hexagon
20. Circle

Drawing Shapes with Specific Attributes

Shape	Drawing	Number of Sides and Vertices	Notes
Circle		0	Not a polygon because there are no sides or angles
Pentagon		5	
Square		4	4
Octagon		8	The shape of a stop sign
Triangle		3	Many different types like equilateral (which has 3 sides of equal length and 3 angles of the same measure), isosceles, scalene, and right triangles
Rectangle		4	A square is a type of rectangle, but not all rectangles are squares

Drawing Shapes with Specific Attributes

Shape	Drawing	Number of Sides and Vertices	Notes
Trapezoid		4	It has two bases
Rhombus		4	A parallelogram with four sides of equal length
Hexagon		6	
Oval		0	Like a circle, it is not a polygon
Parallelogram		4	A square is a type of rectangle, but not all rectangles are squares
Septagon /Heptagon		7	

Classifying Triangles

1. Equilateral

2. Scalene

3. Isosceles

4. Scalene

5. Isosceles

6. Isosceles

7. Equilateral

8. Scalene

9. Scalene

10. Isosceles

11. Equilateral

12. Scalene

13. Scalene

14. Scalene

15. Isosceles

Classifying Triangles

1. Acute

2. Acute

3. Acute

4. Acute

5. Right

6. Obtuse

7. Acute

8. Obtuse

9. Acute

10. Acute

11. Acute

12. Acute

13. Acute

14. Acute

15. Acute

Classifying Triangles

1. Scalene, right

2. Equilateral, acute

3. Scalene, right

4. Scalene, right

5. Scalene, acute

6. Isosceles, acute

7. Isosceles, acute

8. Scalene, acute

9. Scalene, acute

10. Scalene, acute

11. Scalene, right

12. Isosceles, acute

13. Scalene right

14. Isosceles, right

15. Scalene, acute